D1488904

**Protection of workers
against noise and vibration in the working environment**

30 5
1 25

International Labor Office

Protection of workers against noise and vibration in the working environment

TD
892
.I59
1977

International Labour Office   Geneva

ISBN 92-2-101709-5

First published 1977

ILO publications can be obtained through major booksellers or ILO local offices in many countries, or direct from ILO Publications, International Labour Office, CH-1211 Geneva 22, Switzerland. A catalogue or list of new publications will be sent free of charge from the above address.

Printed by Icobulle Imprimeur SA, Bulle, Switzerland

Contents

Introduction

At its 192nd Session, early in 1974, the Governing Body of the International Labour Office decided to convene a meeting of experts on noise and vibration at the workplace as part of the Organisation's activities aimed at improving the working environment. The meeting duly took place,[1] and the experts adopted the present code of practice, publication of which was approved by the Governing Body in March 1975.

This code provides guidance for governments, employers and workers. It sets out the principles that should be followed for the control of workplace noise and vibration, and contains the information required for the establishment of control programmes for individual plants.

It has no compulsory character; it does not lay down minimum requirements but is rather intended to stimulate, guide and promote noise control at the national level. The technical standards that it lays down are objectives which can be attained by successive stages.

---

[1] At the International Vocational and Technical Training Centre in Turin, from 2 to 10 December 1974. It was attended by the following experts:

Dr. A. Darabont (Romania), Chief of the Noise and Vibration Laboratory, Scientific Research Institute for Occupational Protection, Bucharest.

Dr. E. Denisov (USSR), Senior Scientific Officer, Noise and Vibration Laboratory, Occupational Safety and Health Institution, Academy of Medical Science, Moscow.

Prof. G. Gerhardsson (Sweden), Industrial Medicine and Hygiene Office, Swedish Employers' Confederation, Stockholm.

Dr. F. Groenewold Alexandry (Mexico), Confederation of Chambers of Industry of the United States of Mexico, General Manager, Phoné S.A., Acoustic consultancy, Mexico City.

Mr. H. O. Hansen (Norway), Secretary, Norwegian Federation of Trade Unions, Oslo.

Mr. L. Heard (Canada), Safety Director, United Steel Workers of America, Canadian Regional Office, Toronto.

Dr. G. Hübner (Federal Republic of Germany), Head of the Laboratory for the Control of Machinery Noise, Siemens A. G., Berlin.

(footnote continued overleaf)

The experts emphasised the importance of noise and vibration control. Noise and vibration were regarded as being two important factors among the many that contributed to the pollution of the working environment, having a detrimental effect on the worker's health and comfort and constituting a burden on the economy in every country. Owing to the growth of industry and transport, there had been a significant increase in the number and power of noise sources over the past two decades. Statistical studies had shown that when noise and vibration exceeded certain thresholds, they impaired health and working capacity; the effects they produced ranged from mere mental and physical inconvenience to severe organic disorders.

From the human point of view, there has been a rapid rise in the number of persons exposed to the deleterious action of noise and vibration. No matter what the causes and the circumstances, noise and vibration are also ultimately responsible for economic losses due to reduced physical and intellectual working capacity, and even for the temporary or permanent elimination from the workforce (through sick leave or early retirement) of many workers affected by occupational disease or accidents caused by noise or vibration.

Dr. Franca Merluzzi (Italy), Lecturer in Occupational Medicine, Italian General Confederation of Labour (CGIL), Milan.

Mr. M. El Meccawi Mustafa (Sudan), President, Sudan Employers' Consultative Association, Khartoum.

Mr. H. J. Schulte (United States), Deputy Assistant Secretary of Labor for Occupational Safety and Health, US Department of Labor, Washington.

Dr. J. M. Vasiliev (USSR), Chief of Laboratory, Central Scientific Research Institute of Labour Protection, All-Union Council of Trade Unions, Moscow.

Dr. G. Wolff-Zurkuhlen (Federal Republic of Germany), Director, Pollution and Radiation Control, Institute for Occupational Protection and Medicine, Karlsruhe; accompanied by Prof. Dr. H. Dupuis, Max Planck Institute for Agricultural Labour and Technology, Bad Kreuznach.

The following international organisations were represented at the meeting: World Health Organisation, International Organization for Standardization, International Electrotechnical Commission, Permanent Commission and International Association on Occupational Health, International Organisation of Employers, International Social Security Association, Commission of the European Communities.

The elimination of noise and vibration at source when buidings, machines and equipment are being designed is fundamental for effective control. As a first step, manufacturers should be required to provide with each machine or piece of equipment that is a potential noise or vibration source a data sheet giving all the necessary information about the level of noise and vibration emitted. Subsequently, maximum noise and vibration levels could be laid down for these items of equipment and it would be desirable for the purchasers to specify the maximum noise and vibration levels for the equipment in question.

Other group control methods are the isolation of noise and vibration sources (by enclosure, by the use of materials that absorb noise and vibration and by location at a distance), and the prevention of the propagation of noise and vibration or the isolation of workers (e.g. on sound-proof premises or anti-vibration platforms). Only when such collective measures cannot be applied should other types of control be used, such as the reduction of exposure duration and the use of personal protective equipment.

The cost of technical safety measures was mentioned on several occasions by some of the experts; others pointed out that it was also necessary to consider the cost of not taking such measures. The experts arrived at the conclusion that the lack of protective measures and supervision was generally more expensive than a suitable safety and medical supervision programme.

The importance of the medical supervision of workers exposed to noise and vibration was emphasised, but the experts drew attention to the shortage of trained personnel and the cost of such supervision. As far as noise was concerned, in particular, it was therefore advisable to start with an audiometric screening examination and then concentrate on any abnormal and pathological findings requiring more detailed medical examination.

Definitions

This code deals with noise and vibration as physical phenomena that may affect the human body and have a deleterious effect on the worker's health and a negative effect on occupational safety.[1]

The words "noise" and "vibration" are in common use and are to be found in any general dictionary; more precise definitions from the physical and physiological points of view may be found in specialised dictionaries and in textbooks of physics and medicine. These terms, like many others, have been defined internationally and nationally. For the purposes of this code the experts, instead of adopting new definitions, have used those already in existence, making suitable reference to the sources used.[2] Asterisks are inserted in the text on the first occurrence of terms included in the following list.

---

[1] General principles of preventive action in these respects are laid down in the proposed Convention and Recommendation concerning the protection of workers against occupational hazards in the working environment due to air pollution, noise and vibration to be considered by the International Labour Conference in the course of its second discussion of the subject, at its 63rd Session, to be held in Geneva in June 1977. Article 3 of the proposed Convention contains the following definitions of "noise" and "vibration" (see ILO: *Working environment: Atmospheric pollution, noise and vibration,* Report IV(2), International Labour Conference, 63rd Session, 1977, p. 56):

"For the purpose of this Convention—

. . . .

"*(b)* the term 'noise' covers all sound which can result in hearing impairment or be harmful to health or otherwise dangerous;

"*(c)* the term 'vibration' covers any vibration which is transmitted to the human body through solid structures and is harmful to health or otherwise dangerous."

[2] All published in Geneva. In addition to the sources specifically cited in this list of definitions, reference should be made to International Standard 1925-1974 of the International Organization for Standardization: *Balancing—Vocabulary.*

4

Audiometer, pure-tone, for general diagnostic purposes [1]

A device using pure tones designed for general diagnostic use and for determining the hearing threshold levels of individuals by *(a)* monaural air-conduction earphone listening, and by *(b)* bone conduction. The apparatus should provide at least eight tones of frequencies: 250, 500, 1 000, 2 000, 3 000, 4 000, 6 000, and 8 000 Hz for air conduction; for bone conduction it should meet the requirements of IEC Publication 177.

——, *pure-tone screening* [2]

A device designed for screening purposes by monaural air-conduction earphone listening using pure tones. The apparatus should meet the specifications of IEC Publication 178, and provide at least five tones of frequencies: 500, 1 000, 2 000, 4 000 and 6 000 Hz.

Decibel A slow response [dB(A)] [3]

Sound level is defined as

$$20 \log_{10} \frac{p_n}{p_0}$$

where p_n is the r.m.s. sound pressure due to the sound being measured, weighted in accordance with the curve A of IEC Publication 179, and p_0 is the reference pressure (2×10^{-5} Pa $= 2 \times 10^{-5}$ N/m^2 $= 2 \times 10^{-4}$ μ bar). This sound level is measured using the precision sound level meter's dynamic characteristic "slow".

---

[1] International Electrotechnical Commission (IEC), Publication 177: *Pure tone audiometers for general diagnostic purposes* (1965), entries 1, 3.1 and 4.1.

[2] IEC Publication 178: *Pure-tone screening audiometers* (1965), entry 3.1.

[3] IEC Publication 179, Second edition, 1973: *Precision sound level meters*, entries 3.2. and 6.7.

Decibel A impulse response [dB(AI)][1]

The impulse-weighted sound level A is defined by

$$L_{AI} = 20 \log_{10} \frac{p_{AI}}{p_o} dB$$

where p_{AI} is the A-weighted sound pressure measured with an apparatus having the characteristics specified in IEC Publication 179A, and p_o is the reference sound pressure: 2×10^{-5} Pa. Impulse sound levels are expressed in decibels (dB) and the weighting used should always be stated, as well as the dynamic characteristic "impulse".

Environment, working[2]

(1) All places of work as well as all the sites and areas where work is carried out including not only the permanent, indoor, stationary places of work which immediately come to mind such as factories, offices, kitchens and shops, but also temporary places of work such as civil engineering sites, open-air places such as fields, forests, roads and oil refineries and mobile ones such as cabs of trucks, seats of tractors and excavators, ships' galleys, flight decks of aircraft, and so on without exception.

(2) Places where workers are found as a consequence of their work (including canteens, and living quarters on board ship).

Frequency band analysis

Noise and vibration analysis using octave, half-octave and third-octave band filters as defined in IEC Publication 225.[3]

---

[1] IEC Publication 179A: *First supplement to Publication 179 (1973): Precision sound level meters: Additional characteristics for the measurement of impulsive sounds,* entry 3.4.

[2] ILO: *Noise and vibration in the working environment,* Occupational Safety and Health Series, No. 33 (1976), p. 4.

[3] *Octave, half-octave and third-octave band filters intended for the analysis of sounds and vibrations* (1966).

Hearing impairment

Hearing loss exceeding a designated criterion (commonly 25 dB, averaged from the threshold levels at 500, 1 000 and 2 000 Hz.).[1] The hearing loss is the difference between the audibility threshold and the standard reference zero at each frequency as defined in International Standard ISO 389-1975.[2]

Infrasound[3]

Acoustic oscillation whose frequency is too low to affect the sense of hearing.

Noise[4]

(1) Any disagreeable or undesired sound.

(2) A class of sounds, generally of a random nature, which do not exhibit clearly defined frequency components.

——, *ambient*

Noise of a measurable intensity which is normally present.

---

[1] International Standard ISO 1999-1975 *(Acoustics—Assessment of occupational noise exposure for hearing conservation purposes)* defines an "impairment of hearing for conversational speech" in the following manner (entry 3.4.): "The hearing of a subject is considered to be impaired if the arithmetic average of the permanent threshold hearing levels of the subject for 500, 1,000 and 2,000 Hz is shifted by 25 dB or more compared with the corresponding average given in ISO 389."

[2] *Acoustics—Standard reference zero for the calibration of pure-tone audiometers.*

[3] IEC Publication 50 (08): *International electrotechnical vocabulary,* Group 08: *Electro-acoustics* (Second edition, 1960), entry 08-05-040.

[4] ibid., entry 08-05-025.

——, *impulsive* [1]

A noise consisting of one or more bursts of sound energy, each of a duration less than about 1 s.

——, *non-steady* [2]

A noise whose level shifts significantly during the period of observation; a distinction is made between fluctuating noise, intermittent noise and impulsive noise.

——, *steady* [3]

A noise with negligibly small fluctuations of level within the period of observation.

Sound, pure [4]

Sound produced by a sinusoidal acoustic oscillation.

—— *level, equivalent continuous* [5]

That sound level—in dB(A)—which, if present for 40 hours in one week, produces the same composite noise exposure index as the various measured sound levels over one week. For the two degrees of noise exposure to be equivalent, it is necessary that, if the sound intensity increases by 3 dB(A), the duration of exposure be reduced by a half.

---

[1] International Standard ISO 2204-1973: *Acoustics—Guide to the measurement of airborne acoustical noise and evaluation of its effects on man*, entry 3.2.2.3.

[2] ibid., entry 3.2.2.

[3] ibid., entry 3.2.1.

[4] IEC Publication 50 (08), op. cit., entry 08-05-015.

[5] International Standard ISO 1999-1975, op. cit., entry 3.3 (including tables 1 and 2).

Ultrasound[1]

Acoustic oscillation whose frequency is too high to affect the sense of hearing.

Vibration[2]

The variation with time of the magnitude of a quantity which is descriptive of the motion or position of a mechanical system, when the magnitude is alternately greater and smaller than some average value or reference.

——, *hand-transmitted*[3]

Intensive vibration can be transmitted from vibrating tools, vibrating machinery or vibrating workpieces to the hands and arms of operators. Such situations occur, for example, in the manufacturing, mining and construction industry when handling pneumatic and electrical hand tools and in forestry work when handling chain saws. These vibrations are transmitted through the hand and arm to the shoulder. Depending on the work situation they can be transmitted to one arm only or to both arms simultaneously. In principle, these hand-transmitted vibrations are in the frequency range of 8-1 000 Hz.

——, *whole-body*[4]

Vibration transmitted to the body as a whole through the supporting surface, namely the feet of a standing man, the buttocks of a seated man or the supporting area of a reclining man. This kind of vibration is usual in vehicles, in vibrating buildings and in the vicinity

---

[1] IEC Publication 50 (08), op. cit., entry 08-05-045.

[2] International Standard ISO 2041-1975: *Vibration and shock—Vocabulary*, entry 2.001.

[3] International Organization for Standardization: *Third draft proposal for guide for the measurement and the evaluation of human exposure to vibration transmitted to the hand*, document ISO/DP 5349, entries 1 and 2.

[4] International Standard ISO 2631-1974: *Guide for the evaluation of human exposure to whole-body vibration, entries 0, 1 and 1 note 2.*

of working machinery. In principle, it applies to vibration transmitted from solid surfaces to the human body in the frequency range 1-80 Hz. Vibration in the frequency range below about 1 Hz is a special problem, associated with symptoms such as kinetosis (motion sickness), which are of a character different from the effects from higher frequency vibrations. The appearance of such symptoms depends on complicated individual factors not simply related to the intensity, frequency or duration of the provocative motion.

1. General

1.1 Duties of employers

1.1.1.   The employer should be responsible for action to reduce by all appropriate means the exposure of workers to noise* and vibration*.[1]

1.1.2.   The employer should be responsible for the organisational arrangements required to prevent the risks due to noise and vibration in the undertaking.

1.1.3.   The employer should establish and publicise (preferably in writing) a general policy emphasising the importance of prevention, and should take the decisions and the practical steps required to give effect to national regulations and to this code of practice.

1.2 Duties of the workers

1.2.1.   (1) The workers should abide by instructions given and recommendations made to them concerning the prevention of noise and vibration.

(2) In particular, workers should—

(a) make use of noise and vibration control devices and techniques;
(b) indicate whenever such devices are faulty or are in need of maintenance;
(c) be willing to undergo the prescribed medical surveillance; and
(d) use the personal protective equipment provided.

1.3. Co-operation

1.3.1.   The employer should secure the workers' co-operation in action to protect their health and to eliminate noise and vibration

---

[1] Asterisks are inserted in the text on the first occurrence of terms included in the preliminary list of definitions.

hazards, and should establish by joint agreement instructions and recommendations for the prevention of noise and vibration.

1.3.2. (1) The employer should co-operate with the workers in devising and implementing programmes for the prevention and control of noise and vibration.

(2) This co-operation should be especially close within any existing joint safety and health committees at the plant level.

1.3.3. Co-operation should be established between manufacturers and buyers of machinery and equipment with a view to reducing the noise and vibration emission of such machines and equipment.

1.4. Inspection by official services

1.4.1. Inspectors called upon to supervise compliance with the regulations should also take into consideration the provisions of this code of practice.

1.4.2. Inspectors should ensure that an effective prevention programme is evolved and put into effect whenever and wherever there is a special risk due to noise or vibration.

1.4.3. (1) Inspectors should attach special importance to proper briefing of workers, and to co-operation between employer and workers in the prevention of noise and vibration.

(2) Inspectors should ensure that the joint safety and health committees, whenever such committees are established, receive the information they need to be effective.

2. Organising principles of prevention

2.1. Aims

2.1.1.   The aim of noise and vibration prevention programmes should be to eliminate those risks or to reduce them to the lowest feasible levels by all appropriate means.

2.1.2.   The noise and vibration to which workers are exposed, and the time during which they are exposed, should not exceed the established limits.

2.2. Control

2.2.1.   Appropriate measures should be taken at the source to prevent generation, transmission, amplification and reverberation of noise and vibration when machinery and equipment is being designed; and noise and vibration levels are factors to be taken into account when machinery and equipment is to be ordered.

2.2.2.   (1) An endeavour should be made to ascertain at which locations, if any, noise or vibration will exceed the established limits.

(2) Such locations should be identified, marked out, and suitably indicated.

2.2.3.   Technical measures should be taken to control noise and vibration with a view to reducing their levels below the maximum permissible levels.

2.2.4.   When this proves impossible, provision should be made by a reorganisation of work, personal protective equipment or any other suitable means to reduce the exposure below the permissible levels.

2.2.5.   The health of workers likely to be exposed to noise or vibration, or both, at levels exceeding the permissible maxima, including workers whose exposure is limited by personal protective equipment or by administrative arrangements which reduce exposure time, should be appropriately supervised.

2.2.6. (1) The monitoring of the working environment* should be systematic, and repeated as often as needed to ensure that noise and vibration risks are kept under control.

(2) Health supervision data should be used to ascertain that the workers involved remain in good health and hence that the prevention programme is achieving its aim.

2.3. Implementation

2.3.1. Every enterprise or department thereof should implement a general prevention programme that takes due account of its own specific features.

2.3.2. (1) Advice for the implementation of a prevention programme should be provided by the safety service, the occupational health service, or an external adviser or body.

(2) The employer should define and assign technical responsibilities in this connection.

2.3.3. If the enterprise is large enough, competent departments, branches or persons with certain responsibilities should have special duties in connection with noise and vibration prevention in—

(a) the design of new buildings and equipment or studies of new processes;

(b) the purchase of machinery or equipment;

(c) contracts entered into with contractors;

(d) the information and training given to workers; and

(e) the purchase of personal protective equipment and the provision of instructions in regard to its use.

2.3.4. Noise and vibration control should preferably be achieved by collective measures with the assistance of a qualified person; improvements that are recommended should be made forthwith by the competent service.

2.3.5.   The personnel responsible for monitoring noise and vibration in the working environment should—

(a) have received appropriate training in the measurement and control of noise and vibration; and

(b) be equipped with suitable instruments.

2.3.6.   The medical supervision of the workers should be carried out—

(a) under the responsibility of a qualified physician competent to interpret the results of the special tests which are made; and

(b) with the assistance of qualified auxiliary staff that has received appropriate training concerning the special tests to be made (including audiometric tests) and the use of personal protective equipment.

2.3.7.   (1) New building, equipment and plants should be designed, and new equipment ordered, with due consideration to the advice given by technically and medically qualified persons.

(2) The service responsible for monitoring the working environment, the medical service and the workers should be kept informed about any change in plant, equipment or process likely to bring about any substantial alteration in the noise and vibration levels.

3. Noise measurement and assessment

3.1. General

3.1.1. Procedures to measure and evaluate noise exposure depend on the goal to be attained. This applies in particular to—

(a) assessment of the risk of hearing impairment*;

(b) assessment of the degree of interference to communications essential for safety purposes; and

(c) assessment of the risk of nervous fatigue, with due consideration to the work to be done.

3.1.2. Noise measurements should be carried out according to standardised methods appropriate for the specific goal and using standards adopted at the international level or their national equivalent.[1]

3.1.3. The provisions of sections 3.2 to 3.4 are useful it—

(a) standards concerning noise and vibration are being prepared; or

(b) doubt arises whether, or in what manner, a certain standard should be used.

3.2. Hearing conservation

3.2.1. Noise measurements should be made in a manner which will show the noise exposure as accurately as is necessary, so that

---

[1] The present national standards for noise measurements are not fully harmonised at the international level. For the same industrial noise level determined for the same purpose, it is possible to obtain different values by applying different national standards. This is the reason why standards adopted at the international level, or such provisions incorporated into national standards, should be preferred. If not, the use of certain limit values can result in workers' being exposed, in their working environment, to different conditions according to the countries concerned. Existing international standards are described in Appendix 1.

the figures obtained may be compared with the noise limits given in paragraph 4.2.2.

3.2.2. When noise is measured, both normal working conditions and conditions involving the highest noise levels should be taken into consideration.

3.2.3. For steady noise*, the sound pressure level at the workplace (work environment) and equivalent continuous acoustic level should be determined in dB(A) according to international and national standards.[1] Frequency analysis[2] should be made in accordance with standardised methods.

3.2.4. (1) For non-steady impulsive noise*,[3] the additional effects of rapid fluctuations should be taken into account by appropriate standardised measurement methods.

(2) In order to assess the actual noise exposure for non-steady, impulsive noise whichever of the following methods gives the higher readings should be used:

(a) measurement with the sound level meter using the impulse response[4] and calculation of the mean value for an eight-hour daily exposure on an energy basis; or

---

[1] International Standard ISO 1999-1975, op. cit., entry 4: "Noise measurements", and entry 5: "Calculation of equivalent continuous sound level for non-impulsive sound that is intermittent or fluctuating". By applying this standard it is possible to determine, first, under entry 4, the sound level for steady noise which is almost unchanged within a week or varies in a regular manner among a few clearly distinguishable levels, and secondly, under entry 5, equivalent continuous sound levels for non-steady intermittent or fluctuating noise.

[2] See preliminary definition of "frequency band analysis".

[3] For non-steady impulsive noise, the actual noise exposure is higher than would be indicated by noise levels measured in accordance with existing international or national standards. Non-steady impulsive noise is said to exist if there is more than 3 dB(A) between the "slow" response and "impulse" readings of an impulse sound meter as defined by IEC Publication 179A.

[4] See "decibel A impulse response" among the preliminary definitions.

(b) the use of a rule by which a certain (positive) correction factor (usually 3 to 10 dB) should be added to the "slow" response[1] values determined in accordance with international or national standards.[2] The value of this correction should depend on the magnitude of the non-steadiness (impulsivity) of the noise to be measured.

(3) Other special measurement methods which are proved to be appropriate should be used for rapid fluctuating noises.

3.3 Oral communications

3.3.1.   Measurements of noise should be made in noisy working areas where—

(a) it is important, for safety reasons, that a worker should be able to hear a message or other signal; or

(b) the worker would be subjected to extra strain, and the work possibly hindered, by difficulties in oral communication.

3.3.2.   Consideration should be given to defining the maximum distance at which speech intelligibility is preserved at normal voice loudness.

3.4. Fatigue

3.4.1.   Measurements of noise should be made in noisy working areas where—

(a) it is important for safety reasons that a worker should not be exposed to extra strain and fatigue resulting from noise; and

---

[1] See "decibel A slow response" among the preliminary definitions.

[2] International Standard ISO 1999-1975, op. cit., entry 6: "Calculation of equivalent continuous sound level of quasi-stable impulsive noise": "For impulsive noise consisting of series of noise bursts of approximately equal amplitudes (for example, noise from rapidly repeated hammering or riveting) an approximation to the partial noise exposure index may be based on a value 10 dB(A) higher than the measured sound level".

(b) the nature of the work performed by the worker is such that the noise is likely to hinder it and to make it more difficult or arduous.

3.4.2.   Maximum noise levels should be established as necessary, with due regard to the work performed.

3.5. Measuring instruments

3.5.1.   The manufacturers of measuring and analysing instruments should provide full information about such instruments and in particular about their use, calibration, maintenance, margins of error and sensitivity, the interpretation of results and accessories.

3.5.2.   Measuring and analysing instruments should be used in accordance with the manufacturer's instructions.

3.5.3.   The measuring and analysing instruments used should meet the relevant international and national standards.

3.6. Instrument accuracy and calibration

3.6.1.   All measuring and analysing equipment should be kept in good condition and calibrated every day it is used. The required calibration equipment should be accurate to within $\pm$ 1 dB.

3.6.2.   Measuring and analysing instruments should be tested at suitable intervals, and a qualified person should complete a certificate of calibration to be kept with the instrument.

3.6.3.   The persons responsible for the maintenance and testing of measuring and analysing instruments should be specially trained, and it should be their responsibility to ensure that those instruments are kept in good condition.

3.7. Recording of data

3.7.1.   When noise is measured in the working environment, adequate data should be collected, especially regarding—

(a) the location, nature, dimensions, and other distinctive features of the place of work where measurements are made;

(b) the source or sources of the noise, the location of the source in the plant and the type of work being performed;

(c) the instrument used, its accessories, the results of calibration tests, and the values indicated;

(d) the location at which measurements were made, and the direction of the microphone;

(e) the number of workers exposed to noise;

(f) the duration of workers' exposure; and

(g) the date and time, and the name of the observer.

3.7.2.   The collected data should be suitably recorded. It would be advisable to have a special form for this purpose.

4. Noise limit levels

4.1. General

4.1.1.   Noise limits should be laid down as a function of the goal to be attained,[1] in particular—

(a) to prevent a risk of hearing impairment;

(b) to prevent interference with communications essential for safety purposes; and

(c) to eliminate nervous fatigue, with due regard to the work to be done.

4.1.2.   The noise limit levels should be reviewed from time to time so as to keep abreast of scientific knowledge, technical developments and possibilities of prevention.

4.2. Hearing impairment

4.2.1.   Depending on the degree of protection wanted, the following limit values should be determined—

(a) a warning limit value that sets the noise level under which there is very little risk of hearing impairment to an unprotected ear for an eight-hour exposure; and

(b) a danger limit value that sets the noise level above which hearing impairment and deafness may result from an eight-hour daily exposure of an unprotected ear.

4.2.2.   In the light of present knowledge, the following values may be recommended—

(a) a warning limit value of 85 dB(A); and

(b) a danger limit value of 90 dB(A).

---

[1] See Appendix 2, and Appendix 1, section 4.

These values are equivalent continuous sound levels* and should be related to noise measurements made in accordance with the relevant procedures outlined in Chapter 3.

4.3. Special provisions

4.3.1. During emergencies, or because of unforeseen technical imperatives, a worker may be temporarily authorised to exceed the daily dose, provided that the next day he recuperates so that the maximum weekly dose (as determined on the basis of paragraph 4.2.2) is respected.

4.3.2. No matter for how short a time, a worker should not, without appropriate ear protection, enter an area in which the noise level is 115 dB(A) or more.

4.3.3. If there are single isolated bursts of noise which can go above 130 dB(A) "Impulse" or 120 dB(A) "Fast", personal protective equipment should be worn.

4.3.4. No worker should enter an area where the noise level exceeds 140 dB(A).

4.4. Ultrasound and infrasound

4.4.1. A survey should be made to find out if any workers are exposed to ultrasound* or infrasound* in their place of work.

4.4.2. Levels of exposure to ultrasound and infrasound should be reduced to and kept at a reasonable value, due account being taken of up-to-date technical information available.[1]

4.5. Oral communications

4.5.1. The noise limits expressed in dB(A), at places of work concerned and for the kind of oral communications envisaged, should be determined with regard to the current technical material available.

---

[1] See Appendix 2 and, for an example of national provisions along these lines on ultrasound, Appendix 3.

4.6. Fatigue and comfort

4.6.1. (1) Hearing conservation should be an important stage in the improvement of the working environment.

(2) The noise levels laid down should be such that work can proceed normally with a minimum of fatigue and discomfort.

(3) In defining these noise levels due account should be taken of the kind of work being done and the available knowledge.

4.6.2. The noise levels determined should ensure adequate comfort, and be considered as objectives to be aimed at.

5. Vibration measurement

5.1. General

5.1.1. Vibration measurements should adequately represent the levels of vibration to which workers are exposed.

5.1.2. Vibration should be measured under standard conditions[1] so that the figures obtained may be comparable with the limits laid down.

5.1.3. (1) Vibration should be measured as close as possible to the point or area through which it is transmitted to the body.

(2) If such transmission has to pass through a cushion or depends on other factors, these factors should be taken into account.

5.2. Measuring instruments

5.2.1. The manufacturers of measuring and analysing instruments should provide full information about such instruments and in particular about their use, calibration, maintenance, margins of error and sensitivity, the interpretation of results and accessories.

5.2.2. Measuring and analysing instruments should be used in accordance with the manufacturer's instructions.

5.2.3. The measuring and analysing instruments used should meet the relevant international and national standards.

5.3. Instrument accuracy and calibration

5.3.1. Measuring and analysing instruments should be suitably calibrated, in accordance with the relevant standards and with the recommendations made about such calibration.

---

[1] Standard methods are proposed at the international level; see Appendix 1. In particular vibration should be measured in all three axes of an orthogonal co-ordinated system.

5.3.2.   Measuring and analysing instruments should be tested at suitable intervals, and the qualified person should draw up a certificate of calibration to be kept with the instrument.

5.3.3.   The person responsible for maintenance and testing of measuring and analysing instruments should be specially trained, and should be responsible for ensuring that the equipment is kept in good condition.

5.4. Recording of data

5.4.1.   When vibration is measured at places of work, adequate data should be collected, especially regarding—

(a) the characteristics of the source of vibration studied and the type of work being performed;

(b) the characteristics of the path or manner in which vibration is transmitted to the human body (whether there are shock-absorbers, cushions, etc.);

(c) the point at which (including the description of any intermediary elements such as seat sheets) and the pick-up device with which measurements were made, and the values obtained;

(d) the instrument used, its accessories and their characteristics (including sensitivity, dynamic properties and fineness of resolution);

(e) the number of workers exposed to vibration;

(f) the duration of workers' exposure; and

(g) the date and time, and the name of the observer.

5.4.2.   The data collected should be suitably recorded. It would be advisable to have a special form for this purpose.

6. Vibration limits

6.1. General

6.1.1.   (1) Vibration limits should be laid down with due consideration to the aim to be achieved and to the degree of protection required[1] especially for—

(a) vibration affecting the hands*[2] and arms (vibrating tools); and

(b) whole-body vibration* transmitted through the supporting surface.

(2) Vibration limits should also be laid down depending on the work to be done and to avoid fatigue.

6.1.2.   The limits should be reviewed from time to time in the light of new scientific knowledge, technical progress and possibilities of prevention.

6.2. Vibration transmitted to hands and arms

6.2.1.   For a continuous exposure, maximum permissible levels of vibration, depending on the daily exposure duration, should be laid down, account being taken of current knowledge.

6.2.2.   When daily exposure to vibration is made up of two or more periods of exposure at different levels of vibration, or when there are adequate regular breaks, different limits may be laid down.

6.3. Whole-body vibration[3]

6.3.1.   In places of work, the limits for whole-body vibration transmitted by the supporting surface should be laid down with con-

---

[1] See Appendix 2, and Appendix 1, section 5.

[2] See "Vibration, hand-transmitted" among the preliminary definitions.

[3] See in particular International Standard ISO 2631-1974. That standard provides guidelines for the prevention of adverse effects on health, performance and

(footnote continued on p. 27)

sideration to the boundary of reduced well-being and proficiency caused by fatigue.

6.3.2.   For a continuous exposure, maximum permissible levels of vibration depending on the daily exposure should be laid down, with due account to the current technical information available.

6.3.3.   When daily exposure is made up of two or more periods of exposure at different levels of vibration, or when adequate regular breaks are allowed, different limits may be laid down.

6.4. Comfort

6.4.1.   It would be advisable to lay down standards ensuring a degree of comfort and to consider these standards as objectives to be aimed at.

---

comfort caused by whole-body vibration. It recommends boundaries for occupational daily exposure of workers in terms of acceleration and as a function of vibration frequency and daily exposure time. The boundaries are different for vibration in the direction of man's longitudinal body axis (head-to-foot) and in the transverse directions (chest-to-back or side-to-side). For each of these conditions three boundaries are specified: *(a)* an exposure limit (health or safety); *(b)* a fatigue-decreased proficiency boundary, above which vibration exposure leads to increased fatigue and potential interference with specific tasks; *(c)* a reduced comfort boundary that should not be exceeded for most persons if discomfort is to be avoided.

7. Identification of risk areas

7.1. Risk assessment

7.1.1. Noise or vibration, or both, should be measured in all places of work where—

(a) the work done or the working environment possibly will involve a risk due to noise or vibration;

(b) occupational safety and health supervision or inspection discloses that such risks may exist; and

(c) the workers maintain that they are subject to an uncomfortable or disturbing level of noise, vibration, or both noise and vibration.

7.1.2. Noise should be measured whenever speech intelligibility is impaired (in a normal voice) at a distance of 50 cm.

7.2. Ambient noise*

7.2.1. A noise survey should be made to ascertain levels of ambient noise in the various shops within the enterprise.

7.2.2. (1) For the purpose of establishing the noise survey, each shop and other place of work should be taken separately.

(2) If necessary, the workplace may be considered as divided up into areas so that the noise characteristics of the working environment could be better determined.

7.2.3. Noise measurements should be made at a distance of about 1.50 m above the floor or work platform and at least 1 metre from walls, and a mean should be taken of the sound levels measured in various directions.

7.2.4. Consideration should be given to the measurement of ambient vibration.

7.3. Sources of noise and vibration

7.3.1. The sources of noise and vibration should be identified by appropriate measurements.

7.3.2. (1) If the figures obtained for noise measurements exceed 85 dB(A),[1] a noise map should be made in places of work.

(2) This map should draw the contours of areas in which noise is equal to or more than 80 dB(A), 85 dB(A), 90 dB(A), 100 dB(A) and 115 dB(A).

(3) These measurements should be repeated on various occasions until exact and proper contours[2] can be established.

7.3.3. If noise or vibration measurements vary widely because of changes in working conditions (as when machine-tool runs unload and then start working) account should be taken of the least favourable conditions, and it may be well to undertake two or more separate series of measurements.

7.4. Assessment of exposure

7.4.1. Measurements should be made at locations normally occupied by the workers in the area under observation.

7.4.2. (1) Measurements of noise should be made in one of the following ways:

(a) either by a measurement made at the level of the worker's head in his ordinary working posture; or

(b) with the microphone at 1 m away from the worker's head in this position, and on both sides; should the figures obtained vary from one place to another, the highest value should be used.

---

[1] Measurements made in accordance with section 3.2.

[2] See Appendix 4.

(2) A supplementary assessment may also be made with an integrating dosimeter of an approved type, checked from time to time; its use should be taught and supervised by a qualified person.

7.4.3.   Measurements should be made of the vibration transmitted to the whole body and of the vibration transmitted to a particular part of the body.

7.5. Other noise measurements

7.5.1.   Measurements of noise levels undertaken in each working environment should also aim at the assessment of noise attributable to the propagation or transmission of noise emitted in adjacent areas.

7.5.2.   Measurements of ambient noise and assessment of noise sources should be supplemented by measurement of reverberation time in certain places, where this measurement is of interest.

7.6. Marking of areas and equipment

7.6.1.   The contours of areas of equal exposure level should be marked.[1]

7.6.2.   Equipment producing noise in excess of the noise limit levels should be clearly marked with an indication of the nature of the risk, and the need to wear personal protective equipment.

7.6.3.   Equipment, fixed or portable, which may cause vibration in excess of the maximum limits laid down should be clearly marked with an indication of the risk, of the maximum period of use and of the personal protective equipment needed.

7.6.4.   (1) Noisy areas within which exposure time should be reduced and workers should wear personal protective equipment

---

[1] See Appendix 5.

should be marked, due account being taken of the measurements made and the contours drawn.

(2) Only workers wearing adequate personal protective equipment should be allowed access to those areas.

(3) A suitable sign, prominently displayed, should forbid entry to all persons not wearing appropriate means of protection.

8. Noise and vibration control: new equipment

8.1. Specifications for new equipment

8.1.1. Manufacturers should so design the equipment they produce (embodying therein suitable devices) that the noise and vibration emitted is at the lowest feasible level.

8.1.2. (1) Manufacturers should provide information concerning the availability of accessories that are not provided with the equipment itself, but that may be useful or essential to control noise and vibration.

(2) They should also provide information concerning the installation of such accessories so as to obtain maximum efficiency.

8.1.3. Manufacturers should provide full information about levels of noise and vibration emitted, as well as on means of controlling them.

8.1.4. When ordering equipment the purchaser should specify maximum limits for the noise and vibration emitted.

8.1.5. The competent authority should establish maximum levels for noise and vibration emitted by all equipment, or by specific types of equipment.

8.2. Testing

8.2.1. Tests to assess noise and vibration should be performed in accordance with internationally or nationally recognised standard procedures.[1]

8.2.2. An analysis by frequency band*[2] of the noise and vibration produced should be made with a view to discovering means of

---

[1] See Appendix 1, section 6.

[2] See "frequency band analysis" among the preliminary definitions.

attenuating that noise or vibration, combating the emission of pure sounds and providing users with the fullest possible guidance.

8.3. Replacement of hazardous equipment or processes

8.3.1.   Whenever possible, processes and equipment producing less noise and vibration should be given preference.

8.3.2.   It should be considered preferable to purchase equipment that is quieter or produces less vibration than to be compelled to take steps against excessive noise and vibration later on.

8.4. Design and installation

8.4.1.   Noise and vibration control should begin with the design and planning of new buildings, installations and processes; it should be based on the relevant technical knowledge, and in particular—

(a) a knowledge of the noise and vibration characteristics of the equipment and processes to be used;

(b) the choice of suitable construction;

(c) the isolation of operations or plant giving rise to high noise or vibration that is difficult to control.

8.4.2.   As far as possible, preference should be given to materials and structures having a high isolating factor or attenuation factor, or both.

8.4.3.   Once suitable equipment has been chosen, its installation should be studied with due regard to—

(a) the kind of noise and vibration likely to be emitted;

(b) the number and type of machines and other equipment;

(c) the number of workers employed on the particular work premises;

(d) the acoustical characteristics of the work premises; and

(e) the noise already present in the working environment.

8.4.4.   Measurements should be made as soon as machines and equipment have been installed, in order to establish the resulting noise and vibration levels.

9. Noise and vibration control in the working environment

9.1. General

9.1.1. Appropriate technical action should be taken to keep noise and vibration in the working environment below the maximum permissible limits.

9.1.2. Should it be impossible to keep below those limits the following action should be taken, in order of preference:

(a) a reduction in exposure time;

(b) the use of personal protective equipment; or

(c) a combination of time reduction and personal protective measures.

9.2. Control methods

9.2.1. (1) The methods used should aim at—

(a) reducing noise and vibration produced and emitted by the sources;

(b) preventing the propagation, amplification and reverberation of noise and vibration; and

(c) isolating the workers.

(2) Noise and vibration should also be attenuated, as appropriate in particular cases, by distance or by isolating the workers liable to exposure, either by collective measures (such as soundproof booths) or by personal protective equipment.

9.2.2. The various control methods may be combined in order to achieve a suitable reduction in noise and vibration levels.

9.3. Control at source

9.3.1. A distinction should be drawn between the following three main kinds of noise and vibration according to their source:

(a) sound or vibration attributable to the vibration of a solid or liquid (mechanical forces);

(b) sound or vibration attributable to turbulence occurring in a gaseous medium (aerodynamic forces); and

(c) sound or vibration attributable to electrodynamic or magnetodynamic forces, or to electrical arc or corona discharge (electrical forces).

9.3.2.   Methods of controlling sources of noise and vibration which should be specifically considered include the following:

(a) reducing the intensity of vibration by dynamic balancing, reducing the driving force acting on the vibrating part, reducing revolutions per minute, and increasing the length of the working cycles;

(b) reducing the emission efficiency of the vibrating parts by increasing their damping capacity and improving the way in which they are attached;

(c) reducing turbulence and the rate of flow of fluids at inlets, in ducts or pipes, and at outlets;

(d) changing from impact action to progressive pressure action;

(e) changing from reciprocating movements to rotating movements;

(f) changing from sudden stopping to progressive braking;

(g) changing from cylindrical toothed gears to helical gears, and from metal gears to gears of other materials if practicable;

(h) design of the shape and speed of tools with due regard to the characteristics of the material worked;

(i) design of adequate systems for fixing the materials or objects to be worked;

(j) prevention of the striking of objects or materials being transported mechanically, and prevention of their free fall from conveying equipment;

(k) design of burners, torches and combustion and explosion chambers with appropriate characteristics;

(l) adequate design of electrical machinery in regard to electrodynamic, magnetodynamic and aerodynamic forces;

(m) insertion of adequate damping joints at connecting points of machinery and equipment;

(n) adequate design of fan blades; and

(o) adequate design of air tubing and ducting systems (compressed air, ventilation air), and gas or liquid tubing systems to prevent the propagation of noise and vibration and resonance build-up.

9.3.3.   Maintenance of equipment should receive special attention so as to prevent any abnormal increase in noise and vibration emitted by the source.

9.3.4.   The maintenance personnel should be adequately trained in oiling, adjusting, replacement of worn-out parts and the regular and correct maintenance of anti-noise and vibration devices.

9.3.5.   When there is more than one type of source of noise in a particular area, the noisiest one should be attended to first if noise is to be effectively reduced.

9.4. Control of propagation, amplification and reverberation

9.4.1.   (1) Since noise and vibration may take different paths from a single source, their transmission should be studied so that it may be controlled with efficiency.

(2) Steps should also be taken to reduce amplification and reverberation.

9.4.2.   Measures to control the propagation of noise and vibration through solids should include especially—

(a) the mounting of machinery on damping foundations isolated from the floor and walls;

(b) the interposition of anti-vibration materials for mounting and floor-joints;

(c) the siting of noisy and vibrating machinery so that it does not come into contact with other parts of the installation and the workroom.

9.4.3.　Measures to control the propagation and reflection of noise through air should include especially—

(a) partially or totally enclosing the source;

(b) using screens and soundproof partitions and linings;

(c) the proper design of premises from the acoustical point of view;

(d) soundproofing of premises (lining the walls, partitions, floors and ceilings with isolating and absorbent materials).

9.4.4.　Measures to control the propagation of noise should include the use of silencers if necessary.

9.4.5.　(1) The acoustical characteristics of the isolating and absorbent materials and the anti-vibration qualities of the material used for the construction of premises, equipment and enclosures should be carefully considered.

(2) The manufacturers should supply detailed information about the noise-transmission and noise-reduction factors of their isolating and absorbent materials.

(3) Users should take this information into consideration in order to choose the most appropriate materials.

9.5. Remote control and isolation

9.5.1.　When equipment emits high levels of noise or vibration, or both, about which little can be done by other means—

(a) the work should be done at a distance by remote control under suitable methods of supervision; or

(b) arrangements should be made whenever possible to install the equipment in remote locations where the least possible number of workers will be exposed to the nuisance.

9.5.2.　When places of work are especially noisy and technical action is not practicable or has proved unsatisfactory, acoustically isolated areas should be provided whenever possible from which all, or the greater part of, the operations required can be undertaken.

9.6. New risks

9.6.1. Action taken to control noise and vibration should not be such as to create new hazards (for instance, the accumulation of flammable gas within a protective enclosure, where an abnormal increase of temperature may also cause a fire).

9.6.2. The means whereby noise and vibration are controlled should be so chosen as not to increase the potential risks involved (for instance, the use of materials likely to absorb dust or oils which may increase the risk of fire).

9.6.3. (1) Heating systems, air conditioning and ventilation systems to control air pollution or to ensure hygienic working conditions should be so designed that they do not increase noise or vibration in the working environment.

(2) Special attention should be given to low frequency vibration and infrasound control.

10. Protective equipment and reduction of exposure time

10.1. General

10.1.1. When noise and vibration levels cannot be brought below the danger limit either by suitable design of equipment or by suitable installation —

(a) the workers should be provided with and have easy access to soundproof booths either totally enclosed (and air-conditioned if necessary), or partially enclosed;

(b) the workers should be provided with anti-vibration working platforms or stands;

(c) the workers should be provided with adequate hearing protection and anti-vibration devices; or

(d) the length of exposure should be limited.

10.1.2. Personal protective equipment and limitation of exposure time should bring workers' exposure within permissible limits.

10.1.3. (1) Personal hearing protection and anti-vibration equipment should on no account be regarded as adequate substitutes for technical prevention.

(2) They are to be used temporarily to keep risks within the limits, until such time as noise and vibration control can be made more effective through technical improvements.

10.1.4. Every effort should be made to ensure that workers actually use the personal protective equipment that is provided.

10.2. Choice of personal protective equipment

10.2.1. Personal protective equipment should afford effective and reliable protection against the risk.

10.2.2. Manufacturers should provide full information about the attenuation and protection afforded by the personal protective equipment marketed by them.

10.2.3. Since the attenuation offered by hearing protectors varies considerably with the frequency, the noise to be reduced should be analysed by band and the attenuation brought about by the devices should be subtracted for each band, and the noise then converted into dB(A).

10.2.4. Provided equivalent protection can be assured, the workers should be free to choose between different kinds and types of personal protective equipment.

10.2.5. The following personal hearing protective equipment should be considered—

(a) earplugs that can be used more than once;

(b) disposable earplugs;

(c) ear muffs; and

(d) helmets and other specialised ear protectors.

10.2.6. (1) Disposal earplugs should be of a type that can be efficiently used.

(2) Earplugs should in any case be worn only on medical advice.

(3) Earplugs made of ordinary cotton wool should be prohibited.

10.2.7. (1) Personal protective equipment should not be uncomfortable or be a source of accidents.

(2) If there is additional risk when surrounding noise cannot be heard, or when communications are difficult, appropriate measures should be taken.

10.3. Testing of protective equipment

10.3.1. Personal protective equipment should be tested for efficiency.

10.3.2. Tests should be carried out according to a method which has been standardised or approved, or both, by the competent authority.

10.3.3. Only such personal protective equipment as has been duly tested or approved, or both, by the competent authorities should be allowed in places of work.

10.4. Issuing and training in use of equipment

10.4.1. Personal protective equipment should be individually provided to the workers, and identified accordingly.

10.4.2. Steps should be taken to ensure that such devices will produce no undesirable effects and will not be especially uncomfortable to use for the worker concerned; this provision applies especially to the use of earplugs.

10.4.3. Workers who normally wear glasses and those who, for safety reasons, have to wear goggles and safety helmets during work, in addition to ear muffs, should receive appropriate equipment.

10.4.4. (1) When individual protective devices are distributed, the need for them, their use and maintenance should be explained.

(2) The requisite instructions should continue to be given from time to time.

10.5. Inspection and maintenance

10.5.1. Personal and other protective equipment against noise and vibration should be inspected periodically to ensure that it has suffered no damage and is in good condition.

10.5.2. Tests should be carried out at suitable intervals to ensure that personal protective equipment remains effective.

10.5.3. A suitable maintenance programme should be instituted, including proper storage of the personal protective equipment when not in use.

10.6. Reduction in exposure time

10.6.1.   When noise or vibration levels cannot be brought within permissible limits, there should be a reduction in exposure time.

10.6.2.   The following possibilities should be considered with a view to reducing the time of exposure —

(a) rotation of jobs;

(b) reorganisation, so that part of the work can be done without exposure to the risks;

(c) provision of breaks during which exposed workers can relax in a quiet environment.

10.7 Co-operation with workers

10.7.1.   Whenever a prevention programme including the use of personal protective equipment or a reduction in exposure time, or both, is implemented, a special effort should be made to secure the co-operation of the workers.

11. Health supervision

11.1. General

11.1.1. The provisions of this chapter set out the objectives to be reached. Their attainment many be achieved progressively at the national level as a function of the local conditions and possibilities.

11.1.2. All workers continuously or occasionally working in areas or in workplaces where noise or vibration exceeds a certain level,[1] especially workers whose protection is ensured by the use of personal protective equipment or by a reduction of exposure time, or both, should, in so far as possible, be subject to appropriate health supervision.

11.1.3. (1) Health supervision may be prescribed for workers exposed to levels of noise or vibration, or both, that are determined by the competent authority and are lower than the maximum permissible limits.

(2) In addition, the occupational health physician should have the possibility of supervising the health of certain groups of workers as required for preventive purposes.

11.2. Organisation and aims[2]

11.2.1. Workers' health supervision should include—
(a) a pre-employment medical examination;
(b) periodical medical examinations; and
(c) medical examinations after sickness or on specific occasions.

11.2.2. The aim of the pre-employment medical examination should be—

---

[1] Which for noise should be the danger limit specified in Chapter 4.

[2] See Appendix 6.

(a) to determine any contra-indication to exposure to noise or vibration, or both;

(b) to detect any abnormal sensitivity to noise or vibration;

(c) to establish a baseline record useful for later medical supervision; and

(d) to advise the workers about the risks they will encounter in their jobs and about the preventive measures to be taken.

11.2.3. The purpose of the periodical medical examinations should be—

(a) to detect the first signs of occupational disease;

(b) to detect the appearance of any abnormal sensitivity to noise or vibration;

(c) to detect signs of stress due to the work or to the conditions of work so that corrective ergonomic action can be taken; and

(d) to continue the task of informing and advising, and ensure that suitable personal protective equipment is being used.

11.2.4. Medical examinations should be carried out after sickness or on specific occasions.

11.2.5. The medical examination performed on cessation of employment should be such as will provide a general picture of the eventual effects of exposure to noise or vibration.

11.2.6. Health supervision should not put the workers to any expense, and medical examinations should as far as possible be undertaken during working hours.

11.2.7. (1) Workers should be informed of the outcome of the medical examinations they have undergone, and should be informed if, in the occupational health physician's opinion, they are suffering from a professional hypo-acousia or from a disorder attributable to noise or to vibration.

(2) In addition, at the workers' request a copy of their medical records should possibly be forwarded to their own doctors.

11.3. Frequency of health examinations

11.3.1. (1) A medical examination should take place on recruitment or before the worker is allotted to a place of work involving exposure to noise or vibration, or both.

(2) Thereafter, periodical examinations should be carried out at intervals to be laid down as a function of the magnitude of the exposure hazard.

11.3.2. Medical examinations should be repeated more often if risks are especially great or if the occupational health physician feels it necessary for the protection of the health of certain workers or groups of workers.

11.4. Structure of the medical examinations

11.4.1. Pre-employment medical examination[1] for noise exposure should comprise—

(a) a case history;

(b) a general clinical examination;

(c) a clinical examination of the ears; and

(d) a screening (or simplified) audiometric test.[2]

11.4.2. Periodical health examinations for noise exposure, which may be carried out by specially trained health personnel, should comprise—

(a) a short case history;

---

[1] One view expressed at the meeting of experts was that in coming years a simplified examination could be carried out by specially trained health personnel; only abnormal cases would then be referred to the physician.

[2] A screening (or simplified) audiometric test is a pure tone audiometric test (transmission through air) carried out separately for each ear at a number of selected frequencies; temporary threshold shift should be avoided. See "Audiometer, pure-tone screening" among the the preliminary definitions.

(b) a simplified clinical examination of the ears; and

(c) a screening (or simplified) audiometric test.[1]

11.4.3. For noise exposure, medical examinations after sickness or on special occasions, as well as medical examinations carried out when hearing impairment is discovered, should include at least—

(a) a case history;

(b) a general clinical examination;

(c) a thorough clinical examination of the ears, nose and throat; and

(d) a complete (or baseline) audiometric test.[1]

11.4.4. For exposure to local vibration transmitted to fingers and hands or to hands and arms the medical examination should comprise—

(a) a case history, with special reference to the specific occupational risk;

(b) a clinical examination; and

(c) special tests which, according to the kind of exposure involved, should explore in particular the vascular system, the skin sensitivity of the hands, and the state of the bones, the joints and the ligaments.

11.4.5. For exposure to whole-body vibration, the medical examination should comprise—

(a) a case history; and

(b) a general clinical examination.

11.4.6. For exposure to noise and exposure to local and whole-body vibration, all supplementary and special medical examinations

---

[1] A complete (or baseline) audiometric test is a pure tone audiometric test conducted in a soundproof chamber, carried out separately for each ear, at the full range of frequencies after a period of a least 12 hours and preferably 16 hours without exposure to noise. See "Audiometer, pure-tone, for general diagnostic purposes" among the preliminary definitions.

should be carried out by the medical specialists concerned when an abnormality is found in the course of the above-mentioned examinations and it requires further investigation.

11.5. Results and interpretation

11.5.1. The results of medical examinations and supplementary examinations and tests of each individual should be recorded in a medical file.

11.5.2. (1) Fitness for any particular job should be certified by a suitable certificate containing no data of a medical nature.

(2) The decision on whether a worker is fit or not for a particular job should be a decision based on all relevant considerations, including the outcome of the medical examinations performed, with due consideration to the working conditions and to the risks encountered in the working environment, as well as to any possible contra-indications.

11.5.3. A decision to the effect that a worker is fit for the job should in certain circumstances be conditional and prescribe certain specific measures.

11.5.4. Pregnant women should not be exposed to vibration.

11.5.5. Special attention should be devoted to young workers exposed to noise and vibration, and to women workers exposed to vibration.

11.6. Audiometric testing

11.6.1. (1) In the case of a maximum permissible audiometric loss, the necessary preventive action and medical treatment should be initiated to limit further loss of hearing on the part of the worker concerned.

(2) In addition, the cause of the hypo-acousia that has been found should be determined through an inquiry carried out at the workplace.

11.7. Audiometric methods and equipment

11.7.1. (1) The audiometer used should meet the relevant international and national standards.[1]

(2) It should have a certificate of calibration and there should be a clear indication of the calibration standards used (zero dB reference).

11.7.2. The audiometer should be maintained and calibrated in accordance with standards approved by the competent authority or in accordance with the guidelines concerning the calibration of such instruments to be provided by the manufacturer.

11.7.3. (1) The room or booth in which audiograms are made should be silent (the noise level within it being lower than 30 dB(A)).

(2) Such a room or booth should be properly ventilated and maintained at a suitable temperature.

11.7.4. Screening audiograms may be carried out with ear muffs.

11.8. Staff training

11.8.1. Audiometric tests should be undertaken by a specially trained staff.

11.8.2. The auxiliary health personnel should be given special training in how to instruct workers in the proper use of their personal protective equipment.

---

[1] See Appendix 1, section 3.

12. Monitoring

12.1. General

12.1.1.   A long-term programme should be drawn up with the workers' participation to keep noise and vibration risks under control. It should include environmental and health monitoring.

12.1.2.   The aims of this programme should be—

(a) to ensure that the preventive action which has been taken is still effective;

(b) to ensure that the levels, as measured previously, remain unchanged or fall;

(c) to ensure that any changes made in manufacturing processes will not lead to the emergence of new risks;

(d) to promote the study and implementation of more efficient preventive measures; and

(e) to ensure that the health of the workers is efficiently protected.

12.2. Environmental monitoring

12.2.1.   Measurements made in places of work should be repeated at suitable intervals, and a record of the readings obtained should be kept in an appropriate manner.

12.2.2.   Measurements made at places of work should be repeated whenever there is a change in premises, equipment or production which might affect noise and vibration. Likewise, they should be repeated with a view to assessing the effectiveness of the preventive action being taken.

12.2.3.   (1) An inspection programme should be drawn up by the enterprise to ascertain whether the technical preventive measures taken remain effective.

(2) This inspection programme should comprise regular periodical inspections, and special checks whenever necessary.

(3) Special attention should be given to the use of an inspection checklist, to the maintenance of equipment, to the technical preventive devices (such as silencers, casings and soundproof barriers) to reduce noise or vibration, or both, as well as to personal protective equipment and its use.

(4) Such inspections should be carried out by competent persons.

12.3. Health monitoring

12.3.1. Statistics should be derived from the data obtained by medical examinations, so as to ascertain whether the preventive action taken has proved satisfactory.

12.3.2. The medical records should also be used for research purposes, and for epidemiological studies intended to help in defining more precisely maximum permissible levels as well as to collect data on individual sensitivity.

12.4. Comparisons of findings

12.4.1. (1) The data provided by noise and vibration measurements in places of work should be correlated with the outcome of medical examinations to ascertain how effective preventive action has been and what groups of workers are most at risk.

(2) Within the programme of an occupational health service, technical and medical dose response records, when properly maintained, should form an important part of an epidemiological long-term follow-up of the working conditions.

12.4.2. The data available within the undertakings should as far as possible be made available for research purposes for the assessment of the noise and vibration hazards and their prevention.

12.4.3. Scientific research institutes in individual countries should undertake, in close co-operation with each other, more intensive research on the biological effects of noise and vibration, and epidemiological studies should be carried out.

Existing international standards and other international provisions

1. WORKING ENVIRONMENT

A number of international labour Conventions and Recommendations deal with the problem of noise and vibration in the working environment. Mention should be made of the following:

(1) Recommendation No. 97 (1953), concerning the protection of the health of workers in places of employment, provides that in such places "all appropriate measures should be taken by the employer to ensure that the general conditions prevailing in places of employment are such as to provide adequate protection of the health of the workers concerned, and in particular that . . . measures are taken to eliminate or to reduce as far as possible noise and vibrations which constitute a danger to the health of workers".

(2) Convention No. 120 (1964), concerning hygiene in commerce and offices, lays down (in Article 18) that "noise and vibrations likely to have harmful effects on workers should be reduced as far as possible by appropriate and practicable measures".

(3) Recommendation No. 120 (1964), concerning hygiene in commerce and offices, lays down (in Paragraph 57(2)) that "particular attention should be paid—

(a) to the substantial reduction of noise and vibrations caused by machinery and sound-producing equipment and devices;

(b) to the enclosure or isolation of sources of noise or vibrations which cannot be reduced;

(c) to the reduction of intensity and duration of sound emissions, including musical emissions; and

(d) to the provision of sound-insulating equipment, where appropriate, to keep the noise of workshops, lifts, conveyors or the street away from offices".

This Recommendation adds (in Paragraph 58) that "if the measures referred to in subparagraph (2) of Paragraph 57 prove to be insufficient to eliminate harmful effects adequately—

(a) workers should be supplied with suitable ear protectors when they are exposed to sound emissions and vibrations likely to produce harmful effects;

(b) workers exposed to sound emissions and vibrations likely to produce harmful effects should be granted regular breaks included in the working hours in premises free of such sound emissions and vibrations;

(c) systems of work distribution or rotation of jobs should be applied where necessary".

(4) Recommendation No. 141 (1970) concerning control of harmful noise in crew accommodation and working spaces on board ship also contains relevant provisions.

The Model Code of Safety Regulations for Industrial Establishments for the Guidance of Governments and Industry and the various codes of practice published by the ILO contain provisions dealing with the control of noise and vibrations, particularly the following:

(1) Regulation 229 of the Model Code of Safety Regulations for Industrial Establishments for the Guidance of Governments and Industry (under revision), deals with ear protection.

(2) The Code of Practice on Safety and Health in Building and Civil Engineering Work lays down (in paragraph 2.7.2) that "noise should be reduced to the appropriate values, which should be fixed by competent authorities" and (in paragraph 2.7.3) that "if noise cannot be rendered harmless, workers should be provided with suitable ear protectors".

(3) The Code of Practice on Safety and Hygiene in Shipbuilding and Ship Repairing specifies (in paragraph 2.8.1) that "the noise of equipment and operations should be kept as low as possible and not exceed 90 dB(A) at any time during work"; and (in paragraphs 2.8.2 to 2.8.4) that "in acquiring and installing new machinery, special attention should be given to noise prevention. If noise cannot be reduced to a safe level, workers should be provided with ear protectors. Workers continuously exposed to noise should undergo periodical medical examinations".

(4) The Code of Practice on Safe Construction and Operation of Tractors contains the following provisions in Section 10 ("Noise"):

"10.0.1. (1) All practicable steps should be taken to reduce the noise associated with the running of the tractor to a level not exceeding the level established by the competent authority. When levels have not been established, the level of 90 dB(A) at the driver's ear is suggested. In all cases the exhaust system should include a silencer.

(2) The noise level indicated in subparagraph (1) should not be exceeded regardless of the mode of operation, the implements used or the presence of a safety cab or frame.

(3) Where it is not feasible to reduce the noise level to 90 dB(A), operators should use ear defenders to reduce the sound levels to within the limits of the table:

Duration per day (hours)	Sound level dB(A) Slow response	Duration per day (hours)	Sound level dB(A) Slow response
8	90	1½	102
6	92	1	105
4	95	½	110
2	100	¼ or less	115

(4) The level of the noise generated by the tractor should be determined by using approved national or international testing methods."

As far as tractor seats are concerned, paragraph 4.2.1 of the Code provides that "Operators' seats should be adequately sprung or suspended to absorb vibrations", and that "particular attention should be paid to vibrations in the 4 Hz to 9 Hz range, which are particularly troublesome".

2. INSTRUMENTS FOR MEASURING NOISE AND VIBRATION

Limit levels cannot be set unless measurement methods are defined. Measurements carried out at the workplace can be validly referred to the established limits only if the measurement has been carried out with standardised equipment and techniques.

Due account should be taken of existing international standards in this field, in particular that of the International Organization for Standardization and those recommended in publications of the International Electrotechnical Commission:

(a) International Standard ISO 266-1975: *Acoustics—Preferred frequencies for measurements*;

(b) IEC Publication 123: *Recommendations for sound level meters*;

(c) IEC Publication 179: *Precision sound level meters* (Second ed., 1973);

(d) IEC Publication 179A: First supplement to Publication 179 (1973): *Precision sound level meters: Additional characteristics for the measurement of impulsive sounds*;

(e) IEC Publication 225: *Octave, half-octave and third-octave band filters intended for the analysis of sounds and vibrations*;

(f) IEC Publication 184: *Methods for specifying the characteristics of electro-mechanical transducers for shock and vibration measurements*;

(g) IEC Publication 222: *Methods for specifying the characteristics of auxiliary equipment for shock and vibration measurement.*

3. AUDIOMETERS

The establishment of a maximum allowable level is based in particular on the results of epidemiological research, comparing sound levels (measured with sound level meters) and the corresponding hearing loss encountered in exposed persons (assessed by means of audiometers). Standardisation of audiometers, their accessories and the reference zero is essential in this connection. It is also necessary if the audiometric examinations carried out are to be reliable and if their results are to be comparable.

Reference should be made to the following international standards concerning audiometers and their accessories:

(a) International Standard ISO 389-1975: *Acoustics—Standard reference zero for the calibration of pure-tone audiometers*;

(b) ISO Recommendation R 226-1961: *Normal equal-loudness contours for pure tones and normal threshold of hearing under free field listening conditions*;

(c) IEC Publication 178: *Pure tone screening audiometers*;

(d) IEC Publication 177: *Pure tone audiometers for general diagnostic purposes*;

(e) IEC Publication 303: *IEC provisional reference coupler for the calibration of earphones used in audiometry*;

(f) IEC Publication 318: *An IEC artificial ear, of the wide-band type, for the calibration of earphones used in audiometry.*

4. CRITERIA FOR NOISE (HEARING AND HEALTH)

Noise exposure standards should take available criteria into consideration. In this connection, reference may be made to a document on environmental health criteria for noise being drafted by the World Health Organisation.

International Standard ISO 2204-1973 (*Acoustics—Guide for the measurement of airborne acoustical noise and evaluation of its effects on man*) classifies the different kinds of noise (steady noise, non-steady noise, fluctuating noise, intermittent noise, impulsive noise, quasi-steady impulsive noise and an isolated burst of sound energy), studies the question of the physical measurements of noise and deals with the evaluation of effects of noise on human beings. This standard considers, in particular, the question of allowable exposure for hearing conversation (with reference to International Standard ISO 1999-1975) and the problem of noise nuisance (with reference to Recommendation ISO R 1996-1971: *Assessment of noise with respect to community response*). Standard ISO 2204-1973 mentions that the question of noise interference with speech communication is being studied with the aim of developing a simple method giving results of practical value.[1]

International Standard ISO 1999-1975: *Acoustics—Assessment of occupational noise exposure for hearing conservation purposes* provides a basis for interested bodies to set limits for tolerable noise exposure during work. Entry 7, note 1, in this Standard states that in many cases the competent authorities have demanded the institution of hearing conservation programmes if an equivalent continuous sound level of 85-90 dB(A) is exceeded.

The hearing impairment criterion defined by International Standard ISO 1999-1975 (entry 3.4) is "impairment of hearing for conversational speech", by which "the hearing of a subject is considered to be impaired if the arithmetic average of the permanent threshold hearing levels of the subject for 500, 1 000 and 2 000 Hz is shifted by 25 dB or more compared with the corresponding average given in" International Standard ISO 389-1975: *Acoustics—Standard reference zero for the calibration of pure-tone audiometers*.

Entry 4 of Standard 1999-1975 specifies that the sound level at the approximate position occupied by the listener's ear should be determined over an appropriate time and expressed in dB(A). The occurring noise levels should be grouped in classes with a width of 5 dB(A) each. In the case of steady noise where the sound level averaged over a short time is almost unchanged within a week or varies in a regular manner among a few clearly distinguishable levels, the measurement may be made with a sound level

---

[1] Technical Report ISO/TR 3352 drawn up by Technical Committee ISO/TC 43—Acoustics.

meter with A-weighting set at slow response. Where there is an impulsive noise component, correction is necessary; according to entry 6, International Standard ISO 1999-1975 is not applicable to impulsive noise consisting of single bursts of noise but can be applied to quasi-stable impulsive noise by using a correction of 10 dB(A) which should be added to the measured sound level.

Entry 3.3 of International Standard ISO 1999-1975 defines the concept of "equivalent continuous sound level" as the "sound level in dB(A) which, if present for 40 hours in one week, produces the same composite noise exposure index as the various measured sound levels over one week". While, for noise with an impulsive component and for quasi-steady impulsive noise, measurements can be made in dB(A) slow response with a suitable correction, or by using another method giving equivalent results, the use of an impulse sound level meter as defined in IEC Publications 179 and 179A has been recommended by the experts. The noise limit levels (warning and danger levels: Chapter 4) are the continuous equivalent noise levels of 85 dB(A) impulse response and 90 dB(A) impulse response;[1] for an exposure of eight hours per day and 40 hours per week to a virtually constant noise level, these values would be respectively 85 dB(AI) and 90 dB(AI).[2]

5. CRITERIA FOR VIBRATION (HEALTH)

As is the case with noise, it is important to measure vibration with standardised equipment and techniques, and workplace exposure standards should be established on the basis of available criteria.[3]

---

[1] In practice, many industrial noises have an impulse sound component and, consequently, the measurement in dB(A) impulse response is often higher (by around 5 dB) than that made in dB(A) slow response. The sound level limits of 85 and 90 dB(A) impulse response recommended would then, in practice, often correspond to the levels currently categorised as being respectively 80 and 85 dB(A), since one generally uses the A-weighting curve and the slow response without always including the necessary corrections.

[2] If the exposure is for four hours per day and 20 hours per week, the noise limit levels would be, respectively, 88 dB(AI) and 93 dB(AI). Whenever the duration of exposure is reduced by a half, the noise level limits are increased by 3 dB(AI).

[3] ILO: *Noise and vibration in the working environment,* op. cit., pp. 4-5, paras. 9, 14.

As far as vibrations are concerned, wide ranging research is still required, especially for simplifying the measurement and assessment methods, before the hazards can be accurately delineated and precise exposure limits established.[1]

In this context reference should be made to the following international documents:

(a) International Standard ISO 2631: *Guide for the evaluation of human exposure to whole-body vibration*;

(b) a draft proposal for an International Standard ISO/DP 5349 for a guide for the evaluation of human exposure to hand-transmitted vibration.

6. NOISE AND VIBRATION CONTROL

Noise and vibration control is subject to a number of basic principles: collective protection should be given preference over personal protection;[2] prevention is more effective if measures are taken at the design stage of machinery and equipment; assessment of noise and vibration emitted by machinery and equipment should be standardised at the international level;[3] finally, noise and vibration emission levels for machinery and equipment should be established.[4]

---

[1] ILO: *Noise and vibration in the working environment,* op. cit., p. 17, para. 97.

[2] "Collective protection methods" are those intended to protect several persons. They are not only applicable to buildings, equipment and work methods, but may also be of a kind used directly by workers, such as specially insulated booths or premises. "Individual means of protection" means all protective measures which, by their nature, are applicable to one single person at a time. ibid., p.7, footnote.

[3] The noise and vibration emission levels indicated by the manufacturers or specified by the purchasers or by the competent authority should be expressed in such a manner that the worker's exposure can be readily deduced from these data. The emission levels which may be expected when the equipment is used under different installation or environmental conditions should be indicated wherever possible.

[4] ILO: *Noise and vibration in the working environment,* op. cit., pp. 6-8, paras. 16, 20, 24, 25.

As far as the assessment of machinery noise is concerned, mention may be made of the following international recommendations:

(a) ISO Recommendation R 495-1966: *General requirements for the preparation of test codes for measuring the noise emitted by machines*;

(b) ISO Recommendation R 1680-1970: *Test code for the measurement of the airborne noise emitted by rotating electrical machinery.*

In the field of noise and vibration control, mention should be made of the following recommendation and standards:

(a) ISO Recommendation R 354-1963: *Measurement of absorption co-efficients in a reverberation room*;

(b) International Standard ISO 1940-1973: *Balance quality of rotating rigid bodies*;

(c) International Standard ISO 2371-1974: *Field balancing equipment— Description and evaluation*;

(d) International Standard ISO 2017-1972: *Vibration and shock— Isolators—Specifying characteristics for mechanical isolation (Guide for selecting and applying resilient devices).*

The application of preventive methods should follow an order of priority: (1) measures taken at the design stage; (2) collective methods; (3) reduction of exposure duration; (4) personal protection.

Reduction of exposure duration is based on the calculation of an equivalent continuous sound level; in this context note should be taken of International Standard ISO 1999-1975 as regards noise exposure and International Standard ISO 2631-1974 as regards exposure to whole-body vibration. To assess the exposure of persons using personal protective equipment, the equivalent continuous sound level is calculated whilst making allowance for the attenuation provided by the personal protective equipment; the data contained in Annex A to International Standard ISO 1999-1975 should be given due consideration here; it is essential to use octave filters that meet the requirements of IEC Publication 225.

Appendix 2

The health hazards of noise, ultrasound, infrasound and vibration

1. NOISE

Noise may be the cause of various types of injury, disorders, annoyance and disturbance. The effects of noise may be physiological, mental and pathological; a distinction is made between the effects on hearing, the effects on other organs of perception and the general effects.

Effects on the auditory system

(a) Masking

Noise of greater intensity may, in certain conditions, mask less intense noises and make them less readily audible, i.e. reduce the subjective intensity. The former are called masking noises and the latter masked noises.

The masking effect may have certain advantages but it may also have severe consequences when it renders less audible or completely inaudible sounds or noise which, in other circumstances, would give warning of an imminent danger against which the subject should seek protection.

(b) Interference with spatial localisation of sound

The presence of intense sound or noise may reduce the capability of localising a sound source in space. In particular, it may become difficult to perceive the movement of the sound source. This may have serious consequences since the closeness of danger may no longer be correctly perceived.

(c) Pain

High levels of sound and noise become difficult to bear and subsequently intolerable; at even higher levels (about 130 dB), auditory sensation is replaced by pain.

(d) Auditory sensation

Auditory sensation does not occur immediately even if the establishment of the sound or noise that produces it is virtually instantaneous; it develops progressively. Some 100 to 200 ms may be required before the maximum level is attained; in the same way when the noise or sound disappears, the auditory sensation terminates only after a short period of time.

If one measures the time between the onset of sound emission and a predetermined voluntary motor response, one encounters what is called the "reaction time". This is relatively long and varies in length depending on the circumstances and a variety of factors such as the sound intensity and the subject and his or her physical and mental status. Even with perfectly healthy and well trained individuals, the reaction time may be between 100 and 400 ms. An increase in the length of the reaction time increases the accident hazard.

(e) Auditory fatigue

Auditory fatigue results in a reduction of auditory sensitivity—a progressive rise in the perception threshold as the duration of noise exposure is increased. The greater the auditory fatigue, the more intense must the noise be to be perceived. Recovery from this fatigue is always complete.

The appearance of auditory fatigue depends not only on the duration but also the intensity of the noise exposure. Noise which is very intense but of a relatively short duration may, for example, cause auditory fatigue equal to that produced by a significantly less intense noise but for which the exposure time is longer.

Auditory fatigue occurs only if the sound or noise level is sufficiently intense (at least 60 or 70 dB); below this level, exposure may be very prolonged without ever causing auditory fatigue.

(f) Pathological effect on hearing

The deleterious effect of noise on hearing is essentially one of permanent hearing loss, also called auditory trauma. This is the result of lesions to the ear caused by very intense noise.

Acoustic trauma is characterised by irreversible hearing loss in a frequency band of varying width. In the majority of cases and, in particular, in hearing damage due to industrial noise, the frequency band affected is centred around 4 000 Hz.

Noise-induced hearing loss does not progress after exposure to the noisy environment is terminated. Consequently, it is only necessary to remove from a noisy environment anybody in whom the onset of permanent hearing damage has been detected, in order to prevent any further aggravation.

Effects on other organs of perception

Exposure to very high noise levels may often disturb the sense of balance and give the impression of "walking in space". Sufficiently high noise levels may also cause vertigo and nausea.

Psychological effects

The psychological effects of noise consist basically of various feelings of discomfort caused by noise. They may occur as malaise and a feeling of discomfort which may extend even to well characterised mental or neurological disorders. Noise may also be the cause of discomfort due to the impairment of speech communication and speech intelligibility and may also reduce intellectual or psychomotor performance.

(a) General psychological disorders

General psychological disorders are widely encountered but, unfortunately, they are usually ill defined and vary from person to person. The cause-and-effect relationship between noise and the observed disorders is not always easy to establish.

They vary widely in type and degree, ranging from malaise to neurological or psychiatric disorders. The malaise may be a feeling of distress, discomfort, annoyance or surprise.

(b) Interference with speech intelligibility

Very intense noise may interfere with speech intelligibility, make it impossible to hold a conversation, follow a lesson, understand a lecture, talk over the telephone, hear recommendations. It may also prevent a warning of imminent danger from being heard.

(c) Interference with intellectual and psychomotor performance

The effects that noise may have on intellectual and psychomotor performance are seen in particular by the greater difficulty subjects experience in a noisy environment when carrying out intellectual operations or tasks which require concentration or special psychomotor ability.

General somatic effects

A distinction is made between two types of physiological reaction to noise: the first is a startle reflex (alarm or stress reaction) as soon as the noise appears; the second is that which develops when exposure to the sound is prolonged.

Prolonged exposure to intense noise leads to fatigue, lassitude sometimes accompanied by general debility. This may be followed by the development of various disorders such as giddiness, fainting, headaches, migraine, loss of appetite, loss of weight and anaemia, depending on the case.

2. INFRASOUND

Infrasound is acoustic oscillation whose frequency is too low to affect the sense of hearing in man. Infrasound has a frequency range from 0 to 20 Hz.

Infrasound has the following effects:

(a) cochleo-vestibular effects: pain which occurs at an intensity of 165 dB at a frequency of 3 Hz, and 140 dB at 15 Hz;

(b) general effects: the appearance of manifestations such as changes in rate of respiration, skin tension, vision disorders in the vicinity of 10 Hz fatigue and somnolence.

3. ULTRASOUND

Ultrasound is acoustic oscillation whose frequency is too high to affect the sense of hearing in man. It has a frequency range above 20 000 Hz.

The majority of biological effects observed in exposure to ultrasound are the result of acoustic energy being converted to heat. Ultrasound is rapidly absorbed in air, and protection against it is very easy. When exposed to high intensity ultrasound, warning is given by a feeling of skin burning.

It should incidentally be noticed that in practice the audible high frequencies which frequently accompany ultrasound are sufficient to cause the effects attributed to ultrasound. It should also be noted that frequently "borderline ultrasonic" noise, i.e. noise in the range of 10 000 to 20 000 Hz, may be a problem for some individuals, particularly for younger persons, even though the noise may not be audible to the persons who have the authority and responsibility for its control.

4. VIBRATION *

The physiological and pathological effects of vibration transmitted to the human body may be grouped in the following four categories:

(1) *Very low-frequency vibration (lower than 1 Hz).* Responsible for the motion sickness which has as its symptoms nausea and vomiting; this sickness is due to the action of changes in acceleration on the labyrinth of the internal ear.

(2) *Low-frequency whole-body vibration (between 1 and 20 Hz).* Gives rise to a wide range of pathological phenomena including lumbago, sciatica in the lumbar region, neckache, hernia and twinges in the intervertebral discs; these manifestations may occur after a period of exposure in persons who were initially in good health.

(3) *Low-frequency vibration (between 10 and 20 Hz) transmitted to the hands and arms.* Leads to high biomechanical and muscular strain as a consequence of resonance.

(4) *Higher frequency vibration (above 20 Hz).* May produce the following pathological effects:

    (a) vibration in the 20-30 Hz range applied to upper limbs is the cause of osseous or articular lesions in the hands, wrists, forearms and upper arms;

    (b) vibration in the 40-300 Hz range produces vascular manifestations; these may be attributed to local disturbance of the nervous system controlling vasomotor action;

    (c) vibration in the 500-1 000 Hz range may cause sensory and trophic manifestions giving a burning or numb sensation.

Appendix 3
Maximum allowable levels of ultrasound[1]

The maximum allowable sound pressure levels at workplaces in the neighbourhood of ultrasound sources should be established in accordance with the data in table 1.

Table 1. Maximum allowable sound pressure levels at the workplace in the vicinity of ultrasound sources

Geometric frequency means by third octave bands (in Hz)	12 500	16 000	20 000 +
Sound pressure level (in dB)	75	85	110

The levels shown in table 1 may be increased in accordance with the data in table 2 when the total duration of ultrasound does not exceed four hours per day.

Table 2. Corrections for sound pressure levels at workplaces in the vicinity of ultrasound sources

Total ultrasound exposure duration (hours)	Correction (dB)	Total ultrasound exposure duration (minutes)	Correction (dB)
1-4	+ 6	5-15	+ 18
¼-1	+ 12	1- 5	+ 24

The duration of ultrasound exposure must be calculated or based on technical documentation.

---

[1] USSR Health Standards—SN 245-71, 313.8, p. 94, "Ultrasound".

Appendix 4
Noise hazard contours

As stated in section 7.3 of the code, noise sources should be sought out and identified; and if the level measured is higher than 85 dB(A),[1] a noise contour study should be carried out.

Sound level measurements should be carried out in the vicinity of a source, i.e. as close as a worker's ear can be brought to it. If in these extreme conditions the sound levels measured are lower than 85 dB(A), it is clearly impossible for a worker to be exposed to an equivalent continuous noise level at the 85 dB(A) warning level (see section 4.2 of the code). If, on the other hand, the sound level measured is higher than 85 dB(A) under the extreme conditions mentioned above, this means that it is possible to have an exposure higher than the warning level and that the hazard sould be controlled. Sound level measurements should be carried out at increasing distances from the source, in all directions, in order to delineate the contours of the zones in which the sound levels are equal to or greater than 80, 85, 90, 100 and 115 dB(A).[1] These measurements should be repeated until contours of sufficient accuracy have been delineated.

The 90 dB(A) contour should be clearly marked since, beyond this line, exposure above the danger level (equivalent continuous noise level of 90 dB(A)) is possible. Workers employed permanently or occasionally in this zone should be duly informed of the hazard and subject to medical supervision (see paragraph 11.1.2 of the code). Technical safety measures should be taken to eliminate this potential hazard zone. If these are not sufficiently effective, steps should be taken to reduce the exposure time so that workers are not exposed to an equivalent sound level of more than 90 dB(A). If this latter measure does not achieve the desired effect, use should be made of personal hearing protective equipment which will reduce the perceived sound level to below the limit laid down by the competent authorities. In addition, comfort is an important aspect of personal hearing protective equipment.[2]

---

[1] These measurements should be carried out in accordance with the requirements of section 3.2 of the code of practice.

[2] The personal protective equipment should be such as to eliminate all hearing hazard. It would in fact be unacceptable for a worker who had worn ear protection for several years to suffer from occupational deafness on account of inadequate personal protection.

The 85 dB(A) contour line should be indicated. Provision may be made for workers permanently or occasionally employed in this zone to be subject to medical supervision—see paragraph 11.1.3. The noise levels should be monitored periodically and suitable records kept.

The "danger level" (equivalent continuous noise level = 90 dB(A)) is either a limit level or a maximum permissible level, but if greater protection is required, the "warning level" (equivalent continuous noise level = 85 dB(A)) may be regarded as the limit level or the maximum permissible level.

Appendix 5
Signs indicating noise zones

The programme of ISO Technical Committee TC 80 (Safety colours and safety signs) includes the study of standard signs, but there is as yet no ISO symbol to indicate noise zones.

As an example, the following signs might be envisaged (see figures 1 to 3):

(1) for the 85 dB(A) zone (warning level): a yellow triangular sign with a black border, located at the external limit of the zone;[1]

(2) for the 90 dB(A) zone (danger level): a blue circular sign containing the silhouette of a head wearing ear protectors outlined in white, the sign being put up at the exterior limit of the zone to indicate that hearing protection measures must be taken; and

(3) for the 140 dB(A) zone: a white circular sign with a red border with an open hand in black, prohibiting access (paragraph 4.3.4 of the code).

---

[1] The contours of the zones would have been determined in accordance with the data given in Appendix 4, and the measurements carried out in accordance with the provisions of section 3.2 of the code of practice.

Fig. 1
85 dB+
Warning sign

black — yellow

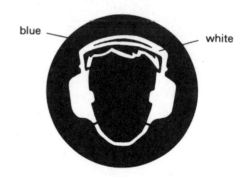

Fig. 2
90 dB+
Hearing protection
must be worn

blue — white

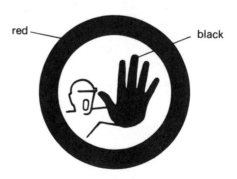

Fig. 3
140 dB+
Proceeding beyond this
sign prohibited

red — black

Appendix 6
Medical aspects: contra-indications

The objectives of the pre-employment medical examination and periodic medical examinations are specified in paragraphs 11.2.2 and 11.2.3 of the code. The result will be an assessment of fitness or unfitness which should be certified in a suitable manner (see paragraph 11.5.2(1)); the decision arrived at should be an over-all one taking into account the results of the medical examinations carried out, the working conditions, the hazards involved in the work and any contra-indications (see paragraph 11.5.2(2)).

1. NOISE

Long-term or permanent contra-indications

There are certain long-term or permanent contra-indications to work in a noisy environment. In particular, unfitness for such work may result from—

(a) a disorder of the hearing or vestibular system, or both; certain chronic diseases of the middle or inner ear; impaired hearing that should not be aggravated, especially if there is increased sensitivity indicating special auditory fragility;

(b) certain clinical diseases, including certain forms of epilepsy and recurrent cephalalgia;

(c) certain mental disorders, namely psychoses and neuroses; contra-indications here should be determined on an individual basis, due account being taken of the whole of the existing health problem and the working conditions.

Temporary contra-indications

In some cases temporary unfitness may be certified for such reasons as—

(a) potentially increased hearing sensitivity to noise due, for example, to streptomycin treatment;

(b) potential increased sensitivity to noise due to a clinical disease, such as rhinopharyngitis or acute middle ear infection;

71

(c) psychological problems such as depression;

(d) the subject's youth (under 18 years of age).

Over-exposure to noise

Certain symptoms of over-exposure to noise demand suitable action—a sensation of muffled hearing, ringing in the ears, and persistence of auditory deficit after the working day.

Audiometry may demonstrate the occurrence of a permanent hearing loss, the progression of which should be halted by suitable preventive measures; the competent authority, in collaboration with suitable specialists, should specify—

(a) audiogram changes which should be regarded as significant in relation to age; and

(b) the various steps to be taken in relation to the measured deficit —increased medical supervision, better personal hearing protective equipment, certification of temporary unfitness or permanent unfitness.

Contra-indications for certain jobs

For certain types of work, good hearing is particularly important and deafness or significant auditory deficit would make the worker unfit for the job. Mention should be made of the following—

(a) jobs in which hearing is an essential safety factor (e.g. professional motor vehicle drivers, crane drivers);

(b) jobs in which hearing is directly related to the work (e.g. telephonist, watch and clock maker, radio technician, automobile or locomotive mechanic), people in contact with the public (e.g. hairdressers, sales staff, beauticians, nurses);

(c) certain jobs with an increased hearing hazard, such as compressed-air workers.

2. VIBRATION

Long-term or permanent contra-indications

There are certain long-term or permanent contra-indications for work entailing exposure to vibration, or more specifically to whole-body vibration, hand-transmitted vibration or both types of vibration. In particular, unfitness for this type of exposure may be attributable to—

(a) disorders of the peripheral nervous system (especially for hand-transmitted vibration): neuritis, polyneuritis;

(b) a disorder of the central nervous system: e.g. sequelae of cranial injury, epilepsy, post-concussion syndrome; or a mental disease: psychosis, neurosis;

(c) cardiovascular diseases: including Raynaud's phenomenon, neurovascular dystonia and angiospasm (in particular for hand-transmitted vibration), stenocardia, hypertension;

(d) gynaecological disorders;

(e) disorders of the urinary system: gallstones, kidney stones;

(f) certain eye disorders, including detachment of the retina and marked myopia;

(g) manifest endocrinological disorders;

(h) organic locomotor system changes, *arthritis deformans* of the hands, and shoulder disorders (in particular for hand-transmitted and arm-transmitted vibration), cervical and spinal arthrosis in general;

(i) miscellaneous diseases such as active pulmonary tuberculosis, stomach and duodenal ulcer, chronic liver disease, ptosis, hernia;

(j) old age (over 50 years).

Temporary contra-indications

The most obvious and most categorical temporary contra-indication is pregnancy, and pregnant women should not be exposed to vibration. Unfitness may also be declared for certain reasons such as—

(a) a recent surgical operation;

(b) certain clinical diseases: chilblains, painful menses, certain gynaecological disorders;

(c) the subject's youth (under 18 years of age).

Over-exposure to vibration

(a) Local vibration

A series of symptoms may indicate the onset and constitute the first stage of a vibration-induced disease—development of paraesthesia, periodic pain in the hands, pronounced whiteness of one or two terminal phalanges of the fingers after exposure to cold water, hyper-aesthesia or hypo-aesthesia—at this stage the disease is still reversible and the symptoms may disappear spontaneously if exposure is terminated.

(b) Whole-body vibration

Disorders indicating over-exposure are more vague and the symptomatology may be related primarily to changes in the central nervous system and the neurovegetative system: there is emotion disturbance, psychasthenia, neurovascular lability, cephalalgia, vertigo; these signs may regress if exposure is terminated.[1]

---

[1] ILO: *Noise and vibration in the working environment,* Occupational Safety and Health Series, No. 33 (Geneva, 1976), p. 84.

How To Have
A Life-Style

How To Have
A Life-Style

QUENTIN CRISP

NEW YORK   LONDON   TORONTO   SYDNEY

March 1980

Manufactured in the United States of America

First American Edition
Published in the United States by
Methuen, Inc.
733 Third Avenue
New York, N.Y. 10017
Designed by Ellen Weiss

Library of Congress Cataloging in Publication Data

Crisp, Quentin.
How to have a life-style.

1. Social history—20th century—Miscellanea.
I. Title.
HN16.C75   1979   158'.1   79-18399
ISBN 0-416-00141-6

AUTHOR'S NOTE

Before I even began to write this book I was aware that there are many people in the world who do not want a style, who will despise the search for one, but since becoming a mail-order guru I have received letters from a number of people who seem to feel that the band is always playing in another street and who seem mysteriously to be locked in their homes. When I tell them that the door is not locked and urge them to go to where the band is playing, they say they haven't a thing to wear. In my opinion, they need wear nothing but their style.

The reason that prompts me to deliver this message is obvious. If I, who am nothing, who am nobody, have been allowed to visit so many of the great cities of the world with my fare paid, then anyone can do it.

What makes this message seem urgent is that the benefits of this way of life are considerable though they may not be what most people imagine. Strangers rush up to me in the street and begin, "Now that you're rich and famous . . . " They never get any further because I stop

them. I am not famous, I am notorious; and if I am rich, it is because I have taken my wages in people.
  They are my reward.

Quentin Crisp
London
June 1979

CONTENTS

How To Have
A Life-Style

BEFORE THE BEGINNING

One day, when I was lying as naked as the Greater London Council would allow on a few planks in the "life" room of Walthamstowe College of Art, a student came and sat beside me. It did not befit my station in life to begin a conversation with her. My supposition was that she wished less to be with me than in front of the only electric heater in the place. I was amazed when she asked me if I would like some of the chocolate that formed the "afters" of her instant lunch.

I sat up at once. My limbs were galvanized, as though insulin had been pumped into my muscles, by the thought of getting something for nothing. The girl broke her slab of chocolate in two and handed me half.

"Oh, not as much as that," I protested. "It would amount to a debauch."

She withdrew her offer altogether. "You talk for talking's sake," she hissed.

I asked if that was bad.

"I mean it," the girl replied. "You talk for talking's sake."

I had heard her the first time and had understood the words but not the contempt with which they were charged. "Would you be equally annoyed," I asked, "if I danced for dancing's sake?"

There was no response because suddenly a member of the staff appeared in the doorway.

Ah, but how wistfully that last sentence reads! It is like a fragment from a costume drama. Today the arrival of an art master in a classroom has no other effect than to add one to the number of hooligans already present. Fifteen years ago his entrance caused guitars to fall silent, grubby limbs to be disentangled, cigarette tins full of cannabis to be shut, and a flimsy pretense of study to fall slowly like a gauze between the student body and mine.

Even though my question was never answered, perhaps never even heard, I wish now I had enlarged it. I should have said, "Would you hate me if I lived for living's sake?" This would have been the total question— the one to which a full reply could have saved the world.

THE NATURE OF STYLE

Style is not the man; it is something better. It is a dizzy, dazzling structure that he erects about himself, using as building materials selected elements from his own character.

Style is the way in which a man *can*, by taking thought, add to his stature. It is the only way.

There is a book written by Mr. Clive Bell on civilization almost the whole bulk of which is taken up with a catalogue of the things that civilization is not. In pursuit of style, I humbly tread a parallel path. Style is not fashion; style is not wealth; style is not learning; style is not beauty. All of this will be demonstrated presently, but it is as well to make the last point clear at the start, because plainness is an excuse so often put forward by failures. A pretty face, they claim, opens all doors. This was never quite true. It may once have opened all bedroom doors, but these have recently been taken off their hinges by the architects of moral open planning. Therefore the keys of this particular kingdom no longer bestow any privilege.

What a stylist needs is glamour—a far more powerful

force than mere prettiness. Glamour exists where something not clearly defined seems to be promised but never given. In any case it is in poetry rather than in the *Times* that we read of society being ruled and empires overthrown by the beauty of woman. It was Mr. Marlowe who believed that Helen burned the topless towers of Ilium; the rest of us may suspect that Mr. Menelaus paid his wife's fare.

Style is an idiom arising spontaneously from the personality but deliberately maintained. It is not stylishness; above all it is not something snatched out of the air and donned to shock or shame the neighbors. Mr. Oscar Wilde, considered by some the great expert on this subject, said that, in matters of importance, it was not sincerity that mattered but style. To the true stylist they are the same thing.

You need to cultivate a life-style first for your own benefit—to give you a firm belief in your own identity and to prevent you from importuning others for their approval to make up for your lack of self-esteem. You need a life-style in your dealings with others only to a lesser degree. It will tell them instantly who and what you are. This cuts the deadwood out of living conversation and makes plain to your friends and even to strangers what questions you can profitably be asked. In the end you have only one thing to offer the world that no one else can give, and that is yourself.

I cannot conceal the agonized contortions by which I am convulsed in my efforts to find a definition of style that will not be too unpalatable to the English. It may

seem strange that this should be so difficult. The soil of
this other Eden used to be ideally suited to the growth of
oddities. Indeed, it has been said that England has more
eccentrics to the square mile than any other country in
Europe, and Dame Edith Sitwell, who made a study of
them, would have agreed. We used to treasure our
eccentrics because they were living examples of the rights
of the individual. With their boyish pranks they defied
the conventions that others accepted but secretly found
irksome. Under a louring social system that made manners more powerful than law and in a landscape where
the garden fence was a barricade, their insouciant follies
seemed like acts of heroism.

In the present permissive society, almost nothing
seems odd. If anything remains to be broken down, it is
the convention of being unconventional. In order to defy
any other unwritten law, a man would have to resort to
acts of the crudest and most wanton bravado. As soon as
we contemplate anything so gross, we realize that, if this
is now the extreme to which eccentrics would be driven,
they would be but weeds in a garden whose flowers are
stylists.

In Britain, sad to say, to make the distinction in this
way is still not necessarily to praise style. There lurk
among us nature lovers who refer to weeds as wild
flowers, dog lovers who prefer mongrels to thorough-
breds, people who would rather do social work than
enjoy society. These men and women are determined to
perpetuate the English ideal of naturalness. That, as a
nation, we have had for centuries so little immediacy

does not for them dim this illusion. Possibly it makes it brighter. Until these last few years spontaneity has never been seen in action on a large scale. Now it has. The full horror of it has been poured over us, but the ideal of behaving naturally still lingers weakly in the hearts of the elderly and the middle-aged. Kept from total collapse only by dint of never calling a looking-glass a mirror, this dying race worships the haphazard, the nonchalant, the unaffected.

There have, of course, been men so perfectly English that they understood this situation and dealt with it by using an anti-style. Such a man was Sir Winston Churchill. Reading his books makes it clear that he was a greater master of prose than of politics, but he was never tempted to let this gift come between him and his audience. His mumbling hesitations and his mispronunciations of foreign (and therefore filthy) words were not blemishes on his style of oratory. On the contrary they were an actor's tricks with which he beguiled the world into believing in his honesty of purpose. How perverse this kind of "methodism" would have sounded to some of his predecessors! Mr. David Lloyd George could never have borne to condescend so far, but then he lived in an age when the huddled masses seldom expected to follow explicitly the whole of any public speech. A dog understands not from words but from a tone of sickening briskness added to its master's voice that it is being invited to take exercise; lovers of literature know that they are listening to poetry because of the reader's swooning diction; so the man in the street of fifty years ago took courage in adversity not so much from strong political argument as from the sight of a mane of white hair unfurled in the wind, from a Welshman's perfect vocal pitch and from orotund references to the Prince of Peace.

Though I secretly long for it, I dare not plead for a return to that world reverberating with rhetoric. I only ask that logic prevail and that we quickly come to realize that, so far from being an obscure insult to our hearers, it is a compliment to them that we present our ideas gift wrapped. People who cry out without premeditation are merely in danger of revealing their emotions—a practice never to be approved. To have feelings is a weakness; to give expression to them is disgusting. Those of us who choose our words carefully stand at least a chance of avoiding this pitfall and, instead, of saying what we mean.

I have been accused of identifying style and standard of living. I don't but I would say this. To a physicist, money would be the solid state of style. That is to say that you can convert your style into riches and enjoy them or you can enjoy your style just the way it is.

As fast as the left-wingers, crouching around their state-owned cauldron, stir the world's wealth into a weak, unappetizing paste, it goes lumpy again. The idealists imagine that, if the planet's resources were doled out fairly, there would be plenty for everyone. In fact, if this were done, prosperity would lie on the earth's crust as thinly as margarine on workhouse bread. This state of affairs would be acceptable to moralists who believe in atonement. Stylists, who prefer happiness, could only deplore it. A modest sufficiency cramps style; extreme poverty, like great danger, enriches it.

Here is an example. In Soho—the hooligan district of London—there lived a woman known to her fellow hooligans as "The Countess." She suffered from almost all the ill effects of undernourishment and exposure to inclement weather. She had no fixed address and no means of support and her body was perpetually bent double from a lifelong habit of looking into trash cans to see if she could find something she could possibly sell to a kind friend or, if not, that she herself might be able to use.

One day, in a garbage can in the most expensive part of London, she found a complete backless, beaded dress. She longed for night to fall so that she could nip into a dark doorway and try it on, but by about half past six her patience was worn out. It was barely dusk but she went into a churchyard in the middle of London and there she proceeded to take off her clothes. This caused a crowd to collect and the crowd caused a policeman to collect.

The next day, in court, when the magistrate said, ". . . and what exactly were you doing, stripping among the dead?" she replied, "I was doing what any woman would be doing at that hour—changing for dinner."

Although there still exists the marvelous world of financial extremes that made this gesture possible, we cannot claim that the atonement-mongers have had no success. They have changed the feel of money; they have made it heavier. Whenever I read about the Medicis—those darlings of the color supplements—I see instantly that this is so. I do not know if Lorenzo called himself "magnificent." Like the earliest governess that I can remember, he may have considered that sort of thing "for others to say." He certainly felt magnificent and seems to have thought of his immense wealth as an attribute of his character. He used it to glorify himself and Florence, which was but a second body. He patronized the goldsmiths and artists of his province not as an act of charity but for the purpose of shameless self-aggrandizement.

Nowadays the rich buy period paintings as an investment and modern ones as a penance.

The nearest thing that I can think of to a modern Medici was Mr. William Randolph Hearst. He was rich beyond the wildest dreams of the tax collectors, and, though money is not in itself style, he made a flat-out attempt to live as though it were. To some extent he succeeded. He took up residence in California, whose climate encourages the growth of personality as Hiroshima favors the spread of panic grass and feverfew. His home there, San Simeon, was the physical territory of his ego. Everyone in the movie colony and some who only wished they were in it struggled to get into his presence. They would allow the most horrible practical jokes to be played upon them simply as a price for being allowed to breathe such opulent air. To modern eyes his image has become slightly confused with that of Mr. Orson Welles, but at the height of his powers he was a colossus of Style.

Journalists who endeavor to belittle him do but nip at his ankles.

If money were all, the rich men who have been lured to California would have gone there presumably for the oil and the aircraft rather than for the movies. To anyone who is purely a financier, the film industry must surely have been an unnecessarily hazardous investment even before the fall. Miss Bessie Love, who made her first film in 1915, says that "Hollywood is just a place—a place where people live," but it is hard not to think that it was also a city of dreams. The lure was the opportunity to enter, at the risk of being stung, a place buzzing with the most beautiful people in the world and take them in the swarm. For a while Mr. Howard Hughes indulged himself with this whim. Then, having invented Miss Jane Russell, he seems to have decided that, in terms of style, there were no new worlds to conquer.

Men like these who in some sense make wealth their life-style are nearly always American or at least Americanized. This is only natural. In the Islands of the Blessed money bestows a kind of sanctity. Elsewhere, far more often than wealth leading to fame, fame becomes a source of income. On the unfashionable side of the Atlantic, wealth has become less an element in which to wallow than something for which the possessor must pay. There is a sense in which, however many noughts we have after our names, conscience has made paupers of us all. The Florentine touch has gone. Although Mr. J. Paul Getty was born in America he became Anglicized. We never saw him, swathed in crimson plush and clanking with amulets, strutting through the supermarkets of Woking.

In a time gone by the rich had many pastimes; the poor had only one. Their numbers were bound to increase. As they multiplied, so did the volume of their lamentations. For centuries the rich ignored the whimpering of the underdogs but finally they became concerned. They did not, I need hardly say, give up their wealth but they changed their style of exhibiting it.

Classicism is the style of pseudopenitence. Mr. Torres, the great designer of clothes for men, expresses the situation sartorially. The French Revolution, he says, killed satins for men.

As long as society, from classical times onward, was built on guiltless wealth and God-given, ennobling poverty, style of a kind grew on trees. We still live in a world where splendor and misery coexist, but the notion that a poor person might, by filling in a coupon or writing a pop song, become rich has robbed the states of both wealth and poverty of their certitude, their continuity, and their sanctity. This makes them poor soil on which to build a style.

It was through the world's treatment of my mother that I personally first became aware that financial barriers were crumbling. My mother was middle-class and middle-brow but she rightly regarded the suburbia in which we were compelled to live as a sort of purgatory of style. She had her own ways of improving the lusterless hour.

There was a moment when, as a child, I stood beside her in a butcher's shop. She was just about to place an order with the man behind the counter when another customer darted into the place and interrupted my mother in order to say, "Mr. Jennings, I simply must tell you that I ate the pork chop you sold me yesterday at ten o'clock at night and slept like an angel." "You'll pardon me," my mother began slowly, "but the angels never sleep. They cry continually day and night—though whether because of the late eating of pork chops the Book of Revelation doesn't say."

When, because it was wartime and various economies had set in, I later in the same day saw my mother lay the fire and black the grate herself, I felt that a discrepancy of style had occurred. It had not. My mother's life-style— that of a muted Lady Bracknell—remained intact but her financial status had undoubtedly changed.

True style changes less and less as it moves toward its perfection and, once complete, is unalterable by outward circumstances or even by time itself.

THE NEED FOR STYLE

It used to be thought that only the rich and famous needed style. Television has changed all that. We can now see that there are people in our society who can earn vast sums of money, become the world's sweetheart, be photographed at airports, and be known by name to hotel proprietors without displaying talent of any kind.

This phenomenon cannot be regarded as an unmixed blessing. Television is certainly to blame for contributing to the madness of young people by bombarding their consciences with the lurid spectacle of world-wide injustice. Even so it would be futile and, worse, styleless, to attempt to limit its activities.

At the very least it should be treated like Chinese rape. As it is inevitable we should relax and enjoy its influence upon our lives. This attitude would at least have the effect of calling the bluff of the newscasters. The spreading of shocking news is not a denunciation of war and similar so-called evils. It is a way of selling television sets. It is a parallel activity with the making of pornographic films. The publicity for such movies is pseudopuritanical

but the audience is invariably pleased to see that debauchery fills the screen for nine-tenths of the film's
showing time, at the end of which retribution gets to
work faster than a clip joint warned that there is going to
be a raid.

An even more stylish method of dealing with rape is
not merely to consent but sexually to assault the rapist. As
Noël Coward has said, television is not a thing to watch.
It is a thing to be on. This is what many people think
but they do not do enough about it. Like Mr. Macbeth
(surely the most styleless Scotsman ever to wear kilts),
what they would highly that would they genteely. This
will not do.

Anyone of us, be he ever so humble, may find himself
"on," and woe to him if he is not prepared with his own
style. He may otherwise find himself simpering and
answering some impertinent interviewer's questions with
"Uh-uh-uh," when in fact he knows, as most of us do,
that he has fascinating opinions about every subject
under the sun.

Some time ago this subject was raised but not really
dealt with in a program called "The Snooper Society," a
documentary feature in which the legality of the invasion
of privacy was questioned.

The inquiry was divided into three parts. The first of
these displayed an assortment of tiny gadgets by means
of which information about our lives can be extracted
from us whether we like it or not. Apparently not only
our telephones but our entire homes can be tapped.
Utterances made in a normal speaking voice as we

wander about the house can be recorded and transcribed in another part of the jungle.

The second section of the program dealt with those personal details about ourselves that we lay before strangers of our own volition. We were shown several private citizens pausing in the street to answer questions about their incomes, their operations and their sex lives. Here it seemed to me that the interviewer spoke from faintly curled lips. In a way this was encouraging. It was an indication that his life-style is at last emerging. At the moment he very likely thinks of himself as an interviewer and regards self-effacement as part of his function. I prophesy that this notion will pass and that soon, like Mr. Muggeridge and others, he will come to treat his subjects as common clay from which to mold a series of busts of himself. What was regrettable about his implied comment (if such it was) was that by it he aligned himself with those who regard indiscretion as a weakness. Surely the people who tell all and worse than all are the saints of census.

Finally, several famous people were asked whether or not they thought the government should be allowed to gather and hoard material for secret files on its victims. One of the celebrities to whom this question was put was an actress, and even she thought that measures should be taken to restrict the authorities. I was deeply sorry to hear her express these views. I like to think of the acting profession as one long lucrative indiscretion. Yet she was not alone in holding this repressive opinion. Mr. Richard Burton and Miss Elizabeth Taylor (who are so great that both of them even brought style to marriage) once complained openly that the newsmen in Italy seemed likely at any moment to photograph one or other of them in the bathroom. The way to deal with this problem (if it is one) is not to build a higher wall around the house but to learn to urinate with style.

If machines come into existence which can collect and tabulate all the relevant facts about everyone in Britain and America and cram them all into a space the size of a telephone booth, it will be done. And what is more, when the information has been gathered, it will be used.

What else has it been gathered for?

By 1984 there will be a governmental and lidless eye gazing down at the occupant of every room in every city in our part of the world. This prospect causes some people to feel positively suicidal. Why? At present many of us spend hours of every evening waltzing round the town without, to our chagrin, causing the slightest stir. When fatigue prevents us from staying out a moment longer, we return home, rip off our eyelashes, and embrace defeat. In a few years' time, however, our anguish will be over. Even at home we shall feel we are worth watching. Margaret or Jimmy will care.

Already, of course, we are being affected by television in the conduct of our daily lives without fully realizing it. Ten years ago a housewife, interrupted in her shopping by a reporter and asked what she thought of the Common Market, would have clapped her hand to her lips, giggled a little, and mumbled that she supposed it was all right. Today it is to be hoped that the same woman in the same situation would spot the camera in a second, smile graciously at it, and, drawing a deep but covert breath, would begin, "While on the one hand certain superficial historians might say . . . "

In this respect the influence of television is entirely good, but we must do more than think of it as an enemy against whose intermittent stares we must brace ourselves. We must welcome its interest and rise joyfully to its challenge by taking our life-style with us wherever we go. It provides us with the ideal means of swatting up our stagecraft. Many confrontations and interviews are stored up for showing at a later date—possibly at several later dates—and this enables us to see ourselves, not, mercifully, as others see us, which is always worse than we really are, but as the camera sees us, which may be a more nearly unbiased view. Thus we have an advantage that was once available only to movie actors who every evening see the rushes of their day's work so that errors can be observed and corrected.

Politicians are already acutely aware that it is far more important to look well on television than to understand world affairs. Mr. Richard Nixon cannot help knowing that the reason he had to wait so many long, dark years before ascending the throne of America was that he allowed himself to be co-starred with his opponent on television. At that moment, whether or not his policies were sound became unimportant. He lost the battle because he lacked that power to project sincerity which Mr. John F. Kennedy, like Dr. Billy Graham and Miss Lena Horne, possessed in such abundance.

Soon there will be very few professions that are not also
the profession of acting. Bishops will bless only that
section of their congregation that is kneeling under the
cameras; surgeons will remove only those organs of the
body that viewers can recognize.

It is the fate of ideas that once seemed contradictory
that they shall one day merge in opposition to some new
concept. Mr. Galileo held that the earth moved around
the sun; the Inquisition said that the sun moved around
the earth. They agreed on only one point, that these two
notions were mutually exclusive—but today an astrono-
mer would probably say that the two planets continually
altered their relative positions.

Once upon a time the dross of everyday existence was
considered quite unsuitable material for drama. Life was
accidental, repetitive, unedifying. Tragedies were the
opposite—formal, articulate and shot through with moral purpose. In spite of the apparent distance between
these two planets they were secretly set on a collision
course from the beginning of time. This first became
noticeable in the work of Mr. Anton Chekhov. We
forgive him because he could not have foreseen the
disastrous consequences of his action, but he was one of
the earliest playwrights drastically to bend the trajectory
of art toward life. Since then things have worsened so
rapidly that in recent years we have come to regard his
plays, which were criticized originally for being "hardly
plays at all," as being too theatrical.

Now the situation is so bad that we can plot the point
where the drama and the documentary will crash. It will

happen on television. In fact it is happening in America already, in the television art form called docudrama.

When the disaster occurs we must decide quite definitely from which vehicle we will drag the victims. Not everyone is of the same mind about this. In the *Guardian* the drama critic Mr. Alan Brien described the effect on his private life produced by appearing in a weekly television program. At first he thought he was gradually bringing to his public image the naturalness of everyday speech. Later he realized that what had happened was the reverse. The formal phrases used by long-term televisionaries were seeping into his conversations with his friends. The article is written with charming self-deprecation but the sentiments expressed therein are nonetheless very worrying. It seems that Mr. Brien prefers the haphazard to the formal—that he is in the ludicrous position of being prepared to take the smooth with the rough. Mr. Kenneth Loach, the television and film director, is even more determined to bring some kind of chaos out of order. He is on record as having said that he wants television plays to look like the news. What a poor, wingless creature he must be! Why does he not want the news to be as well acted as a play?

This, whether they merely watch the news or take part in it, is the dearest wish of all stylists.

We have seen that television greatly increases the need for style. Other aspects of our age, however, make an individual style increasingly difficult to attain. Indeed, to say that is not enough. Our enemy is not so much any one particular period of history as time itself.

The arrow of time points always in the direction of diminishing difference; this was Dr. Jacob Bronowski's first law. Style stands facing the other way; this is mine.

Perhaps time takes an unexpected lurch forward with every major war. It was after the Second World War that I saw the seemingly everlasting walls that were the divisions between countries bulldozed into the dust.

There appeared to be three main reasons why these had to go. In the first place news began to travel so far, so fast, and so incessantly that abroad lost its threatening mystery. In England the aesthetes and the intellectuals had always suffered from a fatuous Francomania, but in 1946, real people started crossing the Channel. Change is ignored by the old, resented by the middle-aged, and welcomed by the young. It was, therefore, chiefly teen-agers who now began to flow over the French landscape like the waters of the Fréjus Dam. They were just about as well received. English campers were shot at by French farmers; tourists in Spain were flung into jail. Increased foreign travel did not bring world peace. All that happened was that, in juxtaposition, the friction between two national styles wore down the finer points of each.

The second force at work to make foreign travel fashionable was the fact that, apart from having the desire for it, the young now also had the money. They became our *nouveaux riches* and, accordingly, were accused by both the traditionally rich and the habitually poor of spending their newly acquired wealth foolishly and with ostentation. Just how this startling financial revolution came about I never discovered. I can only tearfully report as evidence that it did in fact that, when I was a student at Battersea Polytechnic, it took four of us a whole term of cheating, wheedling, saving, and borrowing to consummate an excursion to Box Hill, a beauty spot not fifty miles from London.

The final result of universal tourism was sexual freedom. Once upon a time there were very definite reasons why girls did not wander over the face of the earth alone. These arch fears of violation no longer exist. For rape to take place somebody has to be unwilling. A modern woman, far from dreading a fate worse than staying at home, feels that whatever confusion the limitations of her vocabulary may cause, she will always be able to communicate with a French truck driver or an Italian motorist in the simplest sign language of all.

When all classes and all nationalities became one big unhappy family, there remained only one broad distinction to go. This was the difference between the sexes. In the past few years even that has been obliterated. To many English people this was the most unsettling social change of all. The craze for "boyishness" which broke out among the flappers of the twenties was but a superficial vicissitude in women's fashion. Only prim vicars and stuffy uncles in *Punch* took it seriously. It in no way foreshadowed the epicene being whom we now see everywhere. (A little surprisingly, not all fashion experts are in favor of this. One, at least, says he will reintroduce the codpiece in an effort to design trousers that the wife simply cannot wear.)

The English of even thirty years ago were full of preconceptions about sex nearly all of which have turned out to be false. Men were assumed to have inherited certain character traits along with their testicles. They were thought to be strong, sexy, and brave almost by nature. We now see that these attributes were not innate but forced upon them by circumstances—chiefly economic.

Before the machines took over, many men could earn a living only by physical labor. If they wished to show that they could support a wife, they must be seen to be strong. Now a man needs only enough strength to press a switch. Muscles are for adornment only. On the other hand a woman who wished to become a wife had once to make a great mystery of the rituals of domesticity but now, as the newspapers put it, marriage is dead. It can at last be freely admitted that a man has no more difficulty than his girl friend in opening a tin.

When Mr. Spencer Tracy asked Miss Katherine Hepburn what she noticed first about a person, she replied, "Whether it's a man or a woman." She must be really worried now.

And we should worry too.

The various social changes that I have just described have, during the past fifteen years, brought into being that classless, stateless, all-purpose human unit that now wanders up and down King's Road or any similar street in any of the big cities of the world. This being has been deprived of all the smaller group styles on which he could have begun to build an individual style. To start from the ground up, without help, seems too arduous a task. He finds it easier to identify merely with the fact of being young and has more in common with the teenagers of Tokyo than with an elder brother. On this idea he leans with all his waning might, but it is too large a concept to build upon; it is rather a dream in which to flounder. The only strength it offers is numerical strength. There is danger in numbers.

In no sense can any of the pop festivals be reasonably regarded as a success—even by those who were there. Only a statistician can be exhilarated by the togetherness of 300,000 people. All an ordinary mortal can do is enjoy the company of the people on his right and left and count the noughts in tomorrow's press reports. As for the music, in pouring rain he will hear it gushing from amplifiers which he cannot control instead of listening to it in comfort on his record player at home.

To the rest of us descriptions of all such events read like news of a hurricane. A few people die; others are rushed to the hospital; total disaster is averted only by flocks of helicopters, herds of ambulances, and regiments of nuns riding to the rescue.

To smile benignly on those few occasions when no one is actually murdered is like visiting a mental hospital and expressing delight at the all-pervading calm without realizing that the inmates are all drowsy with Luminal. At White Lake, the site of the first great rock concert, marijuana was sold like ice cream or popcorn. In any case what a standard to set for an evening's entertainment! It is as though, asked about a performance of *Aida*, one were to reply, "It was marvelous. I got away without a scratch."

The young have become like those primitive tribes whose oneness is such that their members can communicate with each other without speech and, if necessary, across hundreds of miles of desert. The faces of explorers blossom into seraphic smiles as they tell us of this phenomenon, but it remains a state of being from which most races have moved away by whatever means they could. In some very remote regions of the world the inhabitants resort to finding some such object as a piece of rock, learning its peculiar markings, and then hiding it where no one else will be able to find it. Only thus can they maintain the idea of their individuality and prevent their thoughts from continually sliding back into the tribal consciousness. Perhaps it was for the same purpose that more sophisticated peoples adopted the custom of marriage. After all, what, until recently, was a woman but a curiously shaped object whose characteristics a man memorized before shutting her in a Georgian-fronted cave?

Now that women no longer gladly submit to this treatment, even stranger devices are being employed by certain people to combat the monotony of standardization. Recently men have started whizzing round the world in boats the size of teacups. Unfortunately the very success of this maneuver has defeated its aim. Almost at once it has become difficult to avoid the rush hour going round Cape Horn.

However farfetched some of these antics may appear,

and whether they succeed or fail, they deserve praise. They are some individual's endeavor to project his style onto the whole world. They involve his entire being and they use up energies that might otherwise become destructive. Those who, on the other hand, allow themselves reluctantly to sink into the faceless limbo toward which time is always drawing us either take drugs or pool their resentment with that of other people in the same predicament until they have accumulated vast stockpiles of hatred.

What can be done? We can either legalize cannabis—perhaps distributing it as part of welfare payments—or we can teach the young how to cultivate a life-style by means less expensive than global sailing.

And what about the opposing team—huddling together in their fear of the young? Could not they, too, be armed with a life-style?

Style is never natural; its nature is that it must be acquired. The finishing touches of style are best self-taught, but the basic exercises that lead to style can be learned from others. If a tutor of any subject does not also teach style, he must at least teach the need for it.

This necessity does not yet seem to be universally recognized, not even in the world of education.

When I was a model, working mainly in art schools in the home counties around London, I spent many enlightening hours in railway carriages stuffed with teachers bound for the same destination as myself. Their discourse made me realize that modern education stands ineffectually trembling in the livid and apocalyptic light of the two-day week. This was not the first time that I found it necessary to revise my original notion of the purpose of education.

When I was a child, I never thought as a child. I subscribed meekly to my parents' idea that a good education was a protracted one. Scholarship seemed to them to be a weapon for use against a hostile world and I often heard my mother nagging my father to equip me with it. Later there was a regrouping of forces. "Operation Pelican" set in. This consisted of their uniting to reproach me with the enormity of my school fees. This happened shortly before I left college.

Then my views on education underwent their first drastic change. The scales fell from my eyes and I saw the whole system as a giant extractor in which the government squeezed from its slaves whatever elemental juices they might contain in the hope of coming across something upon which it could feed.

Things have changed once again since then. I now realize that education is a last wild effort on the part of the authorities to prevent an overdose of leisure from driving the world mad. Learning is no longer an improver; it is merely the most expensive time filler the world has ever known.

Most adult males have not yet begun to feel the rigors of the two-day week. Indeed their lust for overtime drives many of them to work more hours a day than their fathers ever did. The new education is therefore showered most lavishly upon women who stay at home and upon adolescents.

The serious press puts forward the view that grown-up women spend their time flicking over the pages of glossy magazines and sighing, wandering from room to room—nay, from alcove to alcove in their open-plan dwellings—and yawning, or looking out of windows worn thin with polishing, and waiting, waiting, waiting. . . .

In spite of the sentimentality of this picture, it is painted with a certain amount of realism. I have known many market research workers and they tell me that householders seldom refuse to see them. Between adding water to the instant *Boeuf Stroganoff* and throwing away the disposable Spode lie acres of unleavened time which housewives are eager to spend telling some complete stranger why they have abandoned one make of washing powder and bought another. Apparently they are so desperate that they have not even stopped to think that the information they are pouring out will fall into the hands of the enemy—the time savers. They seem unaware that they themselves are helping to design more machinery and manufacture more products that will leave them with more hours on their hands for entertaining more researchers.

When one first contemplates this particular aspect of the problem of leisure, one cannot imagine why housewives do not simply bake a longer cake or wash clothes by hand, but this would not be a satisfactory answer. Such activity is lonely, boring, and even uncomfortable. Moreover it plunges us into the William Morris error. Pursued to its limits such a policy would bring back the homespun, the hand-done, a thousand and one maddening archaic affectations. Any amount of neurosis, any depth of Bovaryism is better than that.

Of course the housewives I met in the art schools never admitted to any of this. They told me that while they draw and paint, vacuum cleaners call to them from half-open cupboards under the stairs and weeds mock them from gravel paths. They said—boasted—that they never had a moment to spare. This is true. They had made sure that it should be true by availing themselves of the government's final solution to the problem of leisure. They had become what students call "part-timers"; they had taken up hobbies.

We must, therefore, force ourselves, if only for a moment, to look into the theory behind taking up hobbies as it relates to style. We do this with some repugnance because it creates a world cluttered with things not very well done, but at first glance there seems to be no other way out of the modern predicament.

Great ingenuity is shown by the authorities in this matter. This I know. In most of the institutes that employed me, the "life" room is on the top floor. (It helps to preserve the *La Bohème* tradition.) To reach my place of work I climbed many flights of stairs and traversed many long corridors. As I went I moved through air that was clamorous with the clash of fencing foils, over floors that shook with the neolithic thump of Morris dancing. These two activities do represent an effort at that total physical coordination that is part of general happiness, but I also passed silent doors labelled wood carving, ceramics, and so on. In other words most of the skills taught in evening institutes must be consumed on the premises. They bestow no expertise that can be flaunted in the process of living. On the contrary, once the school doors have shut behind the students, feelings of loneliness and inadequacy return and, by contrast, strike with greater force.

Evening classes are of use only if you learn to sing and dance, because these are activities the results of which the student takes out into the world with him and wears like a crown. People who have learned to sing, whether they ever sing professionally or not, will always have richer, rounder voices; people who have learned to dance will always have bigger, bolder movements—but of what use are pottery and basket weaving? Once the doors of the evening institute clang behind you, you're where you started. If on the way home you were to become involved in an argument with a stranger at a bus stop, you might find yourself saying, "Well, I can't express myself; you'll have to come and see my baskets."

The cure for freedom is not a greater variety of hobbies. These are but anesthetics; they merely suppress the symptoms of derangement. Hobbies are instead of life. What is needed is a greater intensity of living—style.

You will never learn anything about style in a classroom.

I have a television set so old that a friend, calling on me while it was on, said, "You're crazy keen about these horror films, aren't you?" What was in fact being shown

was a party political broadcast. I have learned to interpret the dim, fractured image presented by my machine
and, alas, can indulge in no confusion. When the picture
on my screen is of a uniformed man prodding, pushing,
or dragging a young person along a public thoroughfare,
I know I am not watching "Columbo," in which menace is
often silken and most threats are oblique. No, I am seeing
the news.

The police have one way of dealing with the young; the
rest of the world has gone to the opposite extreme. Mr.
H.H. Munro, who as Saki gave short-story readers much
pleasure early in this century, pointed out that a soft
answer does not turn away wrath but, rather, inflames
anger. This advice has not been heeded.

Apart from annoying students, leniency gives them a
false set of values. "We have our rights," they will say.
Well, first of all I don't think anyone has any rights. I
think you fall out of your mother's womb; you crawl
across open country under fire; you grab at what you
want, and if you don't get it you go without; and you flop
into your grave. So you have to make up your mind
whether to grab what you want, fight for it, or ask for it,
because otherwise you will lose.

There is a place in London called the London School
of Economics. The students there wanted something or
other and caused such a riot that one of the janitors died.
The press said to one of the leaders of this rabble, "Have
you thought of using force to get what you want?" And
the student said, "Haven't ruled out the idea." Now this is
absolute nonsense. This is relying on the benignancy of

precisely the order of things that you are criticizing. If force were used on both sides, a square mile around the London School of Economics would be evacuated, the army would be called in, and *some* policemen and *some* innocent bystanders and *some* soldiers might be killed; but *all* the students would die.

Almost always what annoys students most is the suggestion that they have too much freedom. Being young they have no historic sense and therefore cannot compare present-day license with the restraint of past decades. "You can't screw girls of fifteen," one of them complained—incidentally using a word that would have got him expelled fifteen years ago. The proposed remedy for their "straitened circumstances" is the revolution. The worrying word here is not "revolution" but "the." It shows that their dream is of an actual day—a particular event whereafter everything will be different. They do not recognize that freedom may march but will never stand still long enough to be measured. Beyond the horizon (sunlit) are other horizons (stormy). Revolt is a perpetual process seldom increasing the amount of liberty in any society but merely pushing it around a bit.

What is so strange is that modern rebels march with their heads in a perpetual "eyes left" position. Their fiery gaze is forever fixed on the Ural Mountains, at the edge of You Know Where, that land without lipstick that no true stylist can bring himself to name. Yet the history of that dreary region is the classic example of the uselessness of violent political action. There it was that those who slaughtered their heartless rulers fell almost imme-

diately into the clenched fists of positively cruel oppres-
sors. This fate was not just a piece of jolly bad luck; it was
an instance of a universal law. Hereditary rulers assume
their subjects are hereditary slaves and are inclined to
treat them with gracious condescension; those who fight
beside you on the barricades know from shared experience exactly what sort of swine you can be and, if they
ever come to power, will handle you accordingly. Hence
the only sound that reaches our ears from behind that
famous curtain is not a cheer of jubilation but the
bleating voice of the weak complaining that they are
being victimized.

I would have thought that devotees of political up-
heaval would rather instance the French Revolution. I
am an inveterate cake eater myself but I would never
deny that, after the decapitation of Miss Norma Shearer,
France became an easier place to live in for a greater
number of people.

This is the most you can ask.

All the golden societies of the past to which historians
point and turn their wistful smiles have had what patience players would call a discard pile. They operated on
two levels, with a slave class who worked, ate, slept, and
died and a leisured class who reclined on one elbow and
spoke. Naturally it is from this latter group that we learn
what life at that time was like. It often makes charming
reading but we can hardly take it to be the whole truth.

There is no such thing as freedom for all. The word
"permissive" cannot logically precede the word "society."
The difference between a mob and a society is order, and

any order is prison for somebody. Men are not even born free. Merely to have parents is an almost intolerable restriction. Nor do men as a rule become free. When they leave home they merely enter a more populous jail. The only thing that can be said for going out into the world is that one's failure to find fulfillment can be blamed on a larger number of people than the two unfortunate beings with whom one started off.

Education meets the crisis by stepping into the student cage, flinging a heap of raw grants onto the sawdust, and retreating hastily. The wish that prompts this action is that these sops will be gobbled up and peace at least temporarily restored. The trouble with this scheme is that the number of subjects that can profitably be learned is fast decreasing. Skills are out. If we shall soon have machines into which you put sheep and they come out cardigans, why teach anyone to knit?

Whatever is done, in the next generation, apart from the computer man, everyone, as the author of *Naked Lunch,* Mr. William Burroughs, has said, will remain permanently angry because he can find nothing to do which he considers important.

Some of the people who favor a policy of indulgence toward the young seek to justify it by claiming that teenagers will one day mature. What hope is there of this? Excessive experience does not lead to maturity nor even to sophistication. On the contrary it is crippling. Athletes agree that it is mild but continual exercise that builds tissue; violent activity breaks it down. As with the body, so with the soul.

The feeling among the young that they are being indulged—that they are being patronized however gravely their elders listen to them—is not by any means new. I would go so far as to describe youth as a desert of boredom lying between the mountains of childish elation and the ocean of adult preoccupations. What causes the present variation of this eternal theme to be played *fortissimo* is not that the opportunities for significant activity are growing less but that, while remaining the same, they appear to be so much more numerous. Nowadays people see so much more going wrong with the world over which they have the same meager control as before.

The world now seems a stunningly ignoble place. It has not really grown all that much worse but appears to have done so because we know so much more about it than we did. Communication, which the class magazines are always telling us we lack, is in fact an epidemic. Everything that happens anywhere is told to everyone in detail, at once and in color and since, as the cliché goes, good news is no news, never a day passes but by proxy we are choked with tear gas, bludgeoned with sticks, robbed of millions, raped at the age of eleven, and starved for a lifetime.

The young have not the means to cure the ills of the world: they can only protest—the more angrily because it can be seen that their action is futile. It is years since marchers stopped impeding the normal flow of traffic between Aldermaston and Trafalgar Square, and still nuclear bombs are tested and still more armaments pile up.

To make matters even more pathetic, half the time the world savers are crying out against situations that are not ills at all. War is the most obvious of these. No less an authority than Sir Edmund Leach, the social anthropologist, states that to live under the perpetual threat of invasion is normal and almost certainly salutary.

We all knew it was normal. Even Walt Disney, the Archdeacon of Twee, did not try to bury this fact beneath a heap of sugar. In none of his nature films was there ever a single chipmunk that ate his lunch in peace. After seeing any of these movies one felt that possibly small animals die as often from ulcers as from enemy activity.

To say that war is salutary is another matter. The statement requires an explanation and this Sir. Edmund is prepared to give on behalf of the majority. It increases the solidarity of nations, he says, and reaffirms boundaries. This last remark is the first flash of style to illumine the landscape in a long time. It is another and much more daring way of saying that without a frame there can be no satisfactory picture.

It also makes clear that to great modern thinkers human life is no longer sacred. We should have known that this would be so. Nothing except diamonds is above the law of scarcity value.

Morality is expediency in a long white dress.

Even the Ten Commandments are a set of survival laws handed down to Mr. Charlton Heston when he was transporting an extremely small band of people across a

very hostile terrain. All moral superiority is founded on
practical superiority. They are one—and this is, I think,
what gives the majority a special gloss. What they do is
"right" in a way that is never quite specified. Right for
whom? Right in what way? Morality changes. Take, the
glib example that murder is forbidden but that war is
noble, heroic. That's a very old idea but it's true. Morality
is changeable and it depends on what is geared to
survival.

Twentieth-century leaders find themselves in the opposite position from Mr. Heston. They rule a concourse
of men occupying standing room only in a world fruitful
to the fringe of vulgarity in some places though less so in
others. Perhaps what makes modern warfare seem unpleasant is not only the deadliness of its weapons but also
the nastiness of its propaganda. If it is absolutely necessary to get rid of people, it should not be necessary to
disparage them. In nobler times a warrior drank the
blood of his adversaries and wore the skulls of his
enemies slung behind him like a university hood in the
hope of becoming pervaded by the glory of the conquered. A present-day warrior should kill no one whose head
he would not be proud to mount on a mahogany shield
and hang in the library.

Whatever the reason, it cannot be denied that the
joyous "Bang-bang" of childhood degenerates somewhere along the line into an out-and-out condemnation
of war by adolescents. To measure the intensity of this
repugnance we have only to recollect that in a town called
Lille the schoolchildren burned themselves to death in

atonement for the wrongs done in Biafra. Almost all young people indulge in a daydream in which they spread out their arms and exclaim, "Take me but save the world," but actually to throw away one's life after the crisis is past can only be possible when the whole world present and to come is intolerable.

Whether rage is directed like a bullet against a particular target or, as more often happens, like buckshot that sprays out generally, the cause is the same. Present-day society, with its constant change, its deafening communication, its almost total loss of definition, has produced a climate of bewilderment so profound that in it personality disintegrates in all but the most persisently self-aware. What we need, therefore, is a reaffirmation of personal boundaries. If to look upward is vain because, to put it politely, You-Know-Who is resting, and to gaze about us is unbearable, where else is there to look but within?

THE RULES AND
STANDARDS OF STYLE

Whenever a journey to the interior is proposed as the way of salvation, in England at any rate, there is an outcry. We have drifted—nay, we have plunged—into a vortex of permissiveness so universal that even the lives of our schoolchildren are one long debauch hardly leaving time for lessons, but at the slightest suggestion of an orgy of self-regard mittened hands are raised in horror. It is forgotten that even the Greatest Stylist of Them All only adjured us to love others as much as we love ourselves.

One reason one would imagine that a philosophy of self-regard might seem worth trying is that a passionate interest in others has proved such a disaster. Altruism is based on a misunderstanding. The opposite of love is not hate. They are the same thing. The antithesis of both these emotions is detachment. Indifference may be a kind of dungeon with barred, lightless windows but detachment is a high tower with windows through which, at an enchanting distance, one can be seen by all. Unfortunately, this lofty state of being is not attained easily even by adults. For adolescents it demands very hard work indeed. Therefore, since a grant for it might be difficult to arrange, the first thing that must be done to persuade young people to attempt a detached way of life is to make it seem worth achieving.

In saying all this I am not hiding in the shadow of Mr. Socrates and merely reiterating that the purpose of education is self-knowledge. This is but "Ordinary" level stuff; at "Advanced" level, the stylist's level, we must learn self-projection.

As soon as it is realized that this is the true aim of all teaching it becomes apparent that most of the subjects at present being taught are beside the point. What we need is not massive grants for the visual arts but encouragement to learn singing, dancing, and a whole syllabus of self-glorifying techniques. Painting is only a rebus of self-expression. Why not learn expression itself?

We have come in recent years to see that life is physical, but we must not because of this come to think that life is only things. We do not need the perfectly designed chair; we want a capacity for relaxation even on a bed of nails. We ought not to waste time constructing a "with-it" telephone booth; we should rather cultivate such perfect diction that we can communicate against all odds. We shall find we already have a golden city when we have all become divine beings.

If to implement any of these proposals would cost a lot of money, I would recommend them with diffidence. This is not the case. With almost no expenditure at all, we can move into a world where all speech will be a kind of literature, every movement a form of dance.

All we need to do is to esteem the freedom to reject as highly as the license to accept; to reform ourselves instead of other people; to be aware of the quality of our experience instead of its quantity; to live for living's sake—with style.

An upper room in the boarding house where I live was once occupied by a young man in perpetual trouble concerning his unemployment compensation. He prepared himself for his annual interview with our local branch of the Ministry of Fear by working himself up into a mood of angry defiance. I felt sure that this was a mistake and warned him that an attitude of willing helplessness might be more effective. Almost instantly he agreed. "I *am* helpless," he said. "Now it's a question of projecting my helplessness." Thus for this young man came the dawn of style, because style, in the broadest sense of all, is consciousness.

Most people are at present content to cherish their mere identity. This is not enough. Our identity is just a group of ill-assorted characteristics that we happen to be born with. Like our fingerprints, if they are noticed at all, they will almost certainly be used against us.

You have to polish up your raw identity into a life-style so that you can barter with the outside world for what you want. This polishing process makes your life so formal that by comparison the life of a Trappist monk is an orgy.

The search for a life-style involves a journey to the interior. This is not altogether a pleasant experience, because you not only have to take stock of what you consider your assets but you also have to take a long look at what your friends call "the trouble with you." Nevertheless, the journey is worth making. Indeed, we might say that the whole purpose of existence is to reconcile the glowing opinion we have of ourselves with the terrible things other people say about us.

If, when you peer into your soul, you find that you are ordinary, then ordinary is what you must remain, but you must be so ordinary that you can imagine someone saying, "Come to my party and bring your humdrum friend," and everyone knowing that he meant you.

"After all, it's for your own good as much as ours" is society's wonderful lie to make more and more outsiders conform. They are exhorted to be no more than reasonable; that is to say they are asked to allow their behavior to be governed by other people's logic. They are urged to think constructively; this means to employ their own energies to build other men's edifices.

The fashionable answer to this threat is to abandon the world altogether, but it is no good dropping out until you know where you are dropping in. As we have seen, merely to fling off all chains is to increase self-doubt.

The real solution is to discard society's rules and standards for your own. However stringent you make these, they will never feel as irksome as those laid upon you by the establishment.

What should the rules and standards of a stylist be?

A famous actress once said that before you can create a style of acting you must first create a life-style. This is wickedly perverse. No one should be urged to improve his mode of living merely to enhance his professional ability on the stage. Describing the director Mr. Ted Mann's method of teaching her to act, Miss Barbara Loden, who is married to Mr. Elia Kazan, says, "He'd make me write on a board everything I'd done for the day. Then he'd try to get me to explain why I'd done it. He said one had to work on one's life as one worked on a part." This seems to me a more logical way of putting things. Though in a way, if we are to be true stylists, these two statements express opposite sets of values, they are alike in that they establish, at the highest level, the connection between acting and living. Once we grasp this idea we see immediately what kind of rules must govern the life of the stylist.

An actor's kingdom is divided into three provinces—the dressing room, the rehearsal room, and the stage. Everything else is unreal. We must partition our lives in exactly the same way. Our homes are sometimes dressing rooms and at other times rehearsal rooms; all the world's our stage. Ideally, when the place where we live is our dressing room we must use it only for preparing to "go on stage" or as an area in which to relax. The search for a life-style will occupy a great deal of your day. It would therefore be wise not to waste time on domestic rituals. It is quite unnecessary to clean the place where you live because, after four years, the dirt doesn't become any worse. It is just a question of not losing your nerve.

In actual fact the only actor I ever knew well was a pantomime dame who endured long waits between brief entrances and filled in the spare time by making ashtrays out of steamed phonograph records, but this I shouldn't mention because it is precisely the sort of activity that anyone working on his life-style must avoid. Time fillers—doodling, knitting, playing patience, daydreaming—are forbidden. Because we must never leave the house underrehearsed, we must lay the most elaborate plans for the future but we should never indulge in fantasies that are incapable of fulfillment. Even the fantasies of others ought to be shunned. If books are read at all, they should never be works of fiction but works of ideas. These will teach us logical thinking, enrich our vocabulary, and deepen our understanding of the structure of language. Novels are to be condemned in the opposite way to drugs. While those who traffic in narcotics are subjected to the utmost rigors of the law, leniency is shown to addicts; book pushers, on the other hand, can sometimes be forgiven; book users never.

Happiness is seldom a relationship between ourselves and others. When Miss Barbra Streisand says that the people who need people are the luckiest people in the world she is being a very funny girl. Even if we hanker only after cozy inanimate things we are likely to have a bad time. Human beings are a commodity as difficult to handle as gelignite. More often happiness is the result of a satisfactory relationship between ourselves and our bodies. The establishment and maintenance of this sense of inner harmony is the proper function of physical culture, and the true object of physical exercise is coordination. The contrary barbaric notions held on this subject by the English seem to be the result of an error in education.

I say this with uncharacteristic diffidence. I have had no reason to go within miles of a public school for more than forty years. All I know of the modern version of these establishments comes from the movie *If*. Though I clung with my long nails to the box office of the cinema near where I live, I was forced by unkind friends to go inside and witness this film. It seemed to me that time had changed nothing.

In the school to which I was sent, great emphasis was laid on field sports. These do not impart grace to the human body; they are crippling. The upper air of business houses in London is clamorous with the grunts of middle-aged men complaining—indeed, boasting with roars of manly laughter—of injuries won on the playing fields of Eton. Even those outdoor games that do not involve being mauled or hit by an opponent seem to cause damage if played continually.

Except for cricket, team sports are violent and fraught with emotion. At school they are said by teachers to rid boys of their sexuality and general aggressiveness. I never met a pupil who agreed that this was so. In fact they encourage feelings of hostility. In the days when England had an empire to build or at least to shore up, there was some justification for the worship of games. At that time physical bravery was an export. Public schools provided their inmates with a doll-house copy of the wide and brutal world into which they would soon be flung.

This situation prevails no more.

England is now a little island off the coast of America. As a nation in the autumn of its life we must guard against falling into the error that besets individuals in the same situation. They are renowned for replacing quiet days at the office with bouts of mowing the lawn or clipping the hedge so ferociously that they bring on heart attacks.

A person who is preparing his life-style should avoid playing even the mildest games if they do not constitute part of that style. For one thing, he might not win, and for another, competition of any kind encourages a man to make comparisons between himself and other people, which is a completely misguided activity of mind.

However, all this said, physical training is important. Indeed, it is the keystone of a happy life. Slow, controlled rituals performed in private—with even the breathing deliberate—are recommended as part of a stylist's training. For a more detailed description of these exercises, I would recommend the reading of books on Yoga were it not for the fact that beyond Hatha Yoga lie other more mysterious strata of this cult. The study of these leads almost inevitably to trouble with the INFINITE. It may be true that preoccupation with time has been the downfall of Western man, but it can also be argued that conjecture about eternity is a waste of time.

The other major item in the syllabus of a school for stylists would be voice training. This seems at present to fare even worse than physical culture at the hands of conventional educators. In most schools devoted to general education it is never taught at all. Yet speech, though not necessarily the most immediate, is our most effective means of self-expression. In ordinary schools all that is done to train children's voices takes the form of a few fitful, usually sarcastic attempts by teachers to shame pupils into more elegant diction. This may only mean that students are urged to adopt the same vocal affectations that strangle the staff. This is not merely insufficient; it is a movement away from style. When Mr. Humphrey Bogart sought advice about his lisp, he was warned in the strongest terms never to let it be taken from him. It was part of his life-style, and it never made him inaudible nor did it prevent him from expressing innumerable shades of meaning.

For the same reasons regional accents and dialects should be preserved. For one thing, open disparagement of them encourages children to upstage their parents, and for another, idiomatic speech enriches the texture of conversation. In debate a discrepancy of styles of language produces a humorous effect at least as stimulating as opposition of viewpoints. The actor Sir Bernard Miles says that everyone should learn the academic version of his native tongue but retain the idiomatic speech in which he was brought up. The latter he calls the language of the heart. In a book of this kind the heart is under suspicion, but it is easy to see what Sir Bernard means and to approve his recommendation.

Poverty of diction occurs not when a whole voice is orchestrated into the key of Lancashire or Devon but

when—usually from laziness—a speaker uses one sound
to express two or more different vowels or groups of
vowels. What we must learn is not a crafty way of glossing
over social giveaways but a carefully thought out technique of voice production. When Miss Elsie Fogerty
ruled the kingdom of language from the Central School
of Speech Training and Dramatic Art, which she founded, every one of her students could recite the first four
lines of one of Mr. Shakespeare's sonnets without pausing for breath. Her pupils could speak for hours without
any audible sign of fatigue. They were also set free by
their training from the vocal monotony that imprisons
nearly all of us. Mr. Goacher, who is a friend of mine, says
that a well-taught actor can touch notes in three octaves
without bursting into song. Why should not everyone
learn to do this? It would hardly be necessary to talk sense
if one spoke well enough. Yet these attainments are not
taught to ordinary mortals. Worse, people are not even
made aware that they lack them. Therefore they never
set about teaching themselves. Hardly any home in
England is without a looking glass, but by no means all of
them are equipped with tape recorders. Even the young
people who do possess them seldom use them as a mirror
for diction.

This then is another rule of style. Everyone who
cannot for any reason go to a school of elocution should
acquire a tape recorder and should spend part of every
day reading onto it some piece of modern prose, playing
it back to himself and noting his errors of speech. If he
learns something by heart and then recites it, he trains

two faculties instead of one. A stylist must maintain a
good memory throughout his life. It is not necessary for
him to be a scholar but it is essential that what little he
knows is always within earshot of his conscious mind.
This is the mental counterpart of physical coordination.

By the time we have performed all these self-improv-
ing rituals, of the hours we spend indoors there will not
be a moment left to mope or dope, to ruminate or to
rebel. As can be imagined we shall be living a life of
considerable self-discipline. How could it be otherwise?
We are being received into the Order of Stylists. Accordingly we need a cell where there can be no unwanted
interruption. It is as well, therefore, that we should think
of ourselves not merely as actors but as stars and, in
consequence, never share our dressing room with anyone.

This brings us to our final rule: *Live Alone.*

The continued propinquity of another human being
cramps the style after a time unless that person is
somebody you think you love. Then the burden becomes
intolerable at once. This may seem to be carrying monas-
ticism to unbearable extremes but dry your tears. What is
frowned upon is cohabitation rather than sex.

If sex were still a private matter, it would require little
or no mention in this book. This, alas, is not the case.
Recently it has been given a great deal of uncoverage in
the press, on television, and in the movies. The people
who raise the loudest and most persistent objection to
this are the moralists. Stylists can never concern themselves with ethics, but they too cannot help forming

opinions about excessive sexual license as about any
other misuse of freedom. In becoming a public pastime
and a topic for incessant conversation sex has not increased its style. Indeed, much of what it formerly
possessed it has lost.

Even the change that has recently overtaken men's
clothing (which is in essence a sexual manifestation) is not
really a movement toward style. There are now more
men than women in England. Therefore, without even
knowing it, adolescent boys obey the appropriate natural
law. They wear brighter plumage. Nevertheless, they do
not take as much care of what used to be meant by their
looks as they once did. They don't need to. In Victorian
times, a gentleman could win the love of a young lady
only by means of what he could show her in the drawing
room—his burning gaze, his heavy moustaches, his
broad (padded) shoulders. Now men wear their clothes
so tight that, if a girl can read the small print, she will
know from fifty yards away with what she is being
threatened. If they are on the menu at all in modern
relationships, faces are for afters.

Instant sex is a time- and labor-saving device, but as
leisure and energy are what we now have to excess, this is
no recommendation. For flavor it will never supersede
the stuff you had to peel and cook. This is one of those
unpleasant truths that the permissive society has brought
to light. We are now all dangerously aware that sexual
intercourse is a bit of a bore. What kept the "divine
woman" lark going for all those long, dark centuries was
not an unquenchable erection but romance. If this had

not been so, how could the troubadours of the Middle Ages have managed to hog all that peak-hour viewing time?

Romance was the style in sex.

Of modern woman we are forced to admit that custom can stale her infinite availability.

I once saw a movie in which Miss Mary Astor promised her daughter a rich full life. She did not give a list of ingredients. I would say they were innocence, wonder, romance, debauchery, indiscretion, and death, but *in that order*.

It used to be said of America that she had passed from barbarism to decadence without ever becoming civilized. I would say that modern adolescents went from innocence to debauchery without ever knowing romance, but while inveighing against the permissive society I would not wish to be thought to be mingling my squeaks with those of Mrs. Mary White Mouse, our local arbiter of television morality. I differ from her in that I do not think there is a pin to choose between innocence and debauchery. I complain merely because I feel that everyone, like a passenger on a luxury liner, should be allowed to work his way through the whole menu even if it makes him sick. The young are debarred from doing this. After decadence there can be no civilization; after debauchery, no romance.

To replace romance we are now offered every possible kind of kinkiness. It is not a satisfactory substitute. What is wrong with pornography is that it is a successful attempt to sell sex for more than it is worth. To give a

peripheral example, let me ask how it is that those
Swedish girls who, with their golden muscles rippling
and their finespun hair afloat, are forever running
naked through the woods, never happen to tread on a
thistle—or worse. If we go deeper into the subject we find
that, if ever we compare with those of other readers our
impressions of a pornographic book, someone always
remarks of the hero, "He must have been inexhaustible."

But most real live men are not inexhaustible. The price
that may have to be paid for sexual activity on a massive
scale includes nervous prostration, syphilis, unwanted
pregnancy, social, emotional, and financial entangle-
ment. None of these items is listed in the catalogue.

Once the hideous truth about the high cost of loving is
out, all the perversities of India cannot sweeten it. They
can only make matters worse.

Far be it from me to advocate a return to "family"
entertainment. The word "family" as used in this phrase
describes a dream unit presided over by a daughter's idea
of a father, kept spick-and-span by a son's idea of a
mother, and romped in (innocently) by a parent's idea of
children. In fact, if someone were to set up a production
in which Miss Bette Davis was directed by Mr. Roman
Polanski, it could not express to the full the pent-up
violence and depravity of a single day in the life of the
average family. On the other hand, pornography is no
more satisfying. As the actor and director Mr. Bryan
Forbes has said, so far from being entertaining, it is
depressing—especially to those who take it most seriously, the sexually inadequate. There is a vision that floats in

shimmering mockery before the bulging eyes of people who suffer from some degree of impotence. It tells them that if only they could have sexual intercourse with a red-haired Japanese girl on a green motorbike on a Wednesday, everything would be different. Then they would experience some immeasurable delight of which circumstances have so far deprived them. Alas, it is a mirage! It is not possible in reality to extort from sexual activity more than your nervous system will stand. Man's reach exceeds his grasp or what is Hades for?

It is especially the young who do not seem to know this. They stagger along the primrose path of pleasure, mistakenly believing that it leads to happiness, until their nerves are frayed to bits. This is partly the reason why there are few people between the ages of fifteen and twenty-five who do not have to be waked every morning with stimulants in order that they may drift through the day on sedatives.

Children overeat. Everyone who has dealt with young children knows that infants cry as often from bewilderment at too wide a choice as they do from frustration. Mr. Peter Fonda says that choice is a form of imprisonment rather than a kind of freedom. It was naughty of him to say this. It is a half-truth. It is not choice but indecision that is inhibiting.

A single glance at the installment-payment racket demonstrates our problem. English and American homes are so full of gear and gimmicks and gadgets that the natives have difficulty hacking a pathway to the front door. When, as so often happens, the so-called sellers

come to reclaim their goods, they remove the spin drier
that the would-be purchaser could have used and leave
the grand piano which she never learned to play but
which was so nicely photographed in the catalogue. In
sex the situation is parallel. Certain magazines have built
up a vast mail-order business in sex. In them every
sensual possibility is listed. If everything is listed, everything will be ordered. If everything is ordered, payments
will lapse. We shall have—we already do have—a nation
of emotional bankrupts. In matters of sex people are
being encouraged to pursue a policy of quick returns
with such ferocity and such persistence that in a single
generation they have transformed the double bed into a
dust bowl of style.

Of course nothing that has here been said about sex
applies to romance. Too much freedom—too much
accessibility—in sex leads to a lack of selectivity and a loss
of style. Courtship, with its seasoning of frustration, was
a way of dealing with this. Of love no good whatever can
be said.

The women's liberation movement clamors busily and
noisily for social, sexual, and financial equality. If only
for this reason, it would be doomed. It is failures who
want equality. A stylist never makes any comparison at all
between himself and other people.

There is, however, another and mortal weakness present in this movement. Nearly all women over the age of
twenty-five whom I have met still believe in eternal love.
They still wish to place their destiny in the hands of
another person, still try to bolster up their wilting self-

esteem by nagging, cajoling, bribing some wretched man for his perpetual and patently reluctant approbation. Until women have struck this way of thinking out of their minds, they can never be liberated. Worse, they can never have style.

When I inveigh against both eternal love and the striving for equality, I am not trying to inhibit anyone's sex life. I merely wish to snatch the straw from every woman's beak and prevent her from nesting. If two people share a territory, ultimately they will be left with only the things about which they disagree. If a man gets up at six every morning and his wife does the same, within a fortnight he will be saying, "Isn't it fortunate; we agree." If, on the other hand, he likes the soap to be on the right side of the wash basin and his wife prefers it on the left, they are heading for a divorce. Their friends will exclaim, "You're getting divorced over a cake of soap!" They do not realize that he had to move it every day.

THE PROJECTION
OF STYLE

Like justice and crossword puzzles, style must not only be done, it must be seen to be done.

The projection of style can be effected by three principal means—our speech, our movements, and our appearance. I would place them in that order of importance. Professor Marshall McLuhan was told by an Indian gentleman that the first thing that aroused his interest in a woman was her odor, then her voice and her walk, and finally her appearance. Evidently the effluvium is the message, but since I would find it difficult to explain to anyone how to stink significantly, I concentrate in this book on the other attributes.

Though it is the least powerful way of communicating style, appearance receives more attention than all the other means put together. Presumably this is because there is more money in it. There has never been any limit to the megatons of make-up a woman can be persuaded to buy nor to the acreage of fabric with which she will drape but not necessarily cover her body.

Fashion magazines are aimed at those who have no

idea who they really are. The articles they contain tell
their readers what to think or, worse, what to say they
think. The photographs help the helpless to disguise
their lack of self-awareness in trappings which really
demonstrate nothing more than that the wearer buys
magazines and can afford to buy a new garment of some
kind every week.

It has been prophesied that *Vogue* will soon be a daily
news sheet. Stop-press fashions will be chalked up on the
placards where we are at present accustomed to seeing
"President Bombs Assassin" or "Drug Addict on Singing
Charge." This will not improve matters. The Mary
Quantum Theory has not worked. I assume she hoped
that, if shops stocked all manner of bits and pieces in all
manner of colors and materials, fashion would cease to
be bondage—would lose its tyrannical power to compel
fat people to squeeze themselves into dresses designed
for the slim or edge little old ladies toward skirts above
their knees. Sad to say, what should have been freedom
of choice has turned out to be confusion of aims. Those
people that *Vogue* calls the not-so-young move slowly. By
the time they have summoned up the courage to try
something new, something newer still has come on to the
market. They lose heart. As for the not-so-old, they are
driven absolutely mad by so much movement, like kittens
in a windy garden. Style can never blossom in this
hurricane of change. Such unsettled weather only panics
people into acquisitiveness.

No one seems to understand that one kind of outfit
must be worn over quite a long period of time before the

movements, the tone of voice, and, above all, the opinions that go with it have been selected and mastered. Sir James Barrie seemed to wear the same seedy black overcoat and Queen Mary to lean on the same umbrella for a lifetime. No one had the nerve to accuse either of them of being outmoded. There is a species of tapir that has worn a white stripe across its back for, perhaps, a million years, but this I agree may be going too far even for a stylist.

It would matter less that people are inclined to wear any old—or rather, any new—thing if other more telling means of self-expression were being employed to raise anyone above the flood level of nonentity. They seldom are.

People search for and cling to the fashions in behavior (protest marches, strikes, hijackings) and in phraseology ("with it," "getting it on") as desperately as to fashions in clothes. In England—perhaps in the whole world (except in Ireland where everyone speaks a kind of blank verse)—the great obstacle that style must overcome is not so much ignorance of words as embarrassment at trying them out oneself and mistrust of them when they are used by others.

People seem to experience this reaction in two degrees—violently in connection with the spoken word and mildly with what is written. When a discrepancy between these two levels of vocabulary is noticed, the advice given is always to settle for less. A man answering an advertisement for a job is warned not to employ in his letter of application phrases that he would not dream of using in an interview—not to write that he is seeking a post if he would say he was looking for work. Better advice would be to admonish him never to utter a sentence which, if written down, would look clumsy, repetitious, unfinished.

It would seem that in the last century the middle and upper classes conversed on this principle. Even little boys appear to have uttered such sentences as "Imagine, my dear brother, my grief upon learning of your misfortune." As all I know of the manners of those times comes from reading novels of the period, I cannot tell whether discourse was really of this high order or whether Mrs. Mary Shelley and other such writers merely wished it were. It is certain, however, that early in the twentieth century the spoken word started to lose its formality. It began to exhibit what some people (as though it were praise) call the human element.

It is a pity we allowed this to happen. If the spoken word is not quickly restored to health, we shall soon be in a very bad way indeed, for the written word is on the way out altogether. In England its decline began with the publication of *Picture Post*. This paper reduced journalism to the level of the strip cartoon and was, only then we did not know it, a kind of still television.

Now that the images have begun to move, their ascendancy over words has become more obvious and it will increase. In the work of Mr. Ray Bradbury, who is more or less in control of our future, even shop fronts are devoid of lettering because consumers a few years hence will no longer be able to read even the simplest phrase.

By that time a book will be a little vulcanite box (in a Penguin paperback color to show to which category of literature it belongs). Proceeding from one corner of this contraption there will be a flexible tube with a mechanical device at the end of it which the "reader" will fit into his ear like a hearing aid. Once in a while he will go to the library and say to the girl behind the counter, "I've come to get my Ian Flemings recharged." Then, on the way home on the bus—I mean the municipal helicopter—as he sits plugged in to whatever book is in his breast pocket, a fatuity born of murmuring sound will pass into his face.

When that day comes literary style as we know it will be obsolete. While books are still a succession of printed sentences, a writer spends half of his total effort trying by means of the careful choice of words and the order in which he arranges them to compensate for the absence of the stress, rubato, and vocal pitch that would prevent his hearers from misunderstanding him if he were speaking to them. The voice is a very persuasive instrument. If an income tax form were to be read to us by Mr. Orson Welles, it would sound like *Wuthering Heights* but sexier and even more tragic.

When books as we know them are gone, the future of the language will rest entirely in the long white hands of style seekers. They must not let a single word die. I wish I could suggest a thousand ways in which this could be done but I know of only two. One is to live a long time and the other is never to allow a new word to be used in one's hearing without, preferably at the time, finding out what it means. Dr. Sitwell said to Mr. Clarkson, "Without our

thesauruses where would we be?" I fear that I find it hard
to go along with this. To me the swatting up of synonyms
seems likely to lead to word-dropping, which is as boring
as name-dropping. It may also lead to the misuse of
words. Few dictionaries give the grammatical construc-
tions into which words fit and almost none deals with
their emotional coloring.

This subtle quality that certain words have over and
above their literal meaning is most evident in what used
to be called bad language. Nearly all these words either
name parts of the body that, until the porn cinemas took
over our cultural centers, were seldom seen in public
places or else describe natural functions not usually
performed in public. When a doctor wishes to refer to
any of these organs or their use, he can easily avoid
shocking or embarrassing his patients. He has a whole
medicine chest full of synonyms which positively reek of
antiseptic. Parents, when speaking to their children
about these matters, also wish to rid the subject of all
suggestion of shame or defilement. They borrow or
invent cozy diminutives that will turn the situation into a
game. One's only regret about this practice is that it leads
certain women to continue to use this nursery language
in what might have been their adult lives.

In a third context the specious authority of the Anglo-
Saxons is invoked and the shortest words are used for the
longest things with the deliberate aim of introducing into
polite conversation an element of crudity. This is where
permissiveness defeats itself and does the cause of invec-
tive a disservice. If these short, sharp words are used too

often they will become blunt and their gritty texture will be smoothed away until they give their users no satisfaction and their hearers no shock. They will cease to be effective cries of rage, contempt, or frustration. Worse, they will lose their word-style—their capacity to create unique, humorous effects.

Parenthetically it is worth noting the deleterious effect that the persistent use of bad language has had on the drama. The permissive society has brought many unpleasant facts to light. One of these is that the greatest playwright who ever lived was not Mr. Shakespeare nor Mr. Sheridan nor Mr. Shaw. He was the Lord Chamberlain, the Royal Censor. Why should anyone bother to write with skill or even with tact when he knows he can get his work put on at the Royal Court Theatre any day if he can but remember to sprinkle the pages of his script with that famous word the broadcasting rights of which belong to Mr. Tynan?

The introduction of new words and phrases which, until they have become accustomed to them, pedants call slang does our language no harm. On the contrary, this enriches it. Without such additions speech could never grow. The people who damage English are those who, like impatient jigsaw puzzle fanatics, force words into situations which they do not fit, disregarding their grammatical structure and their emotional overtones.

Much time, energy, and money are spent saving from the rising waters of the Aswan Dam horned animals whose existence is justified only by the help their names give to makers of crossword puzzles. It is a pity so little is done to rescue words from the engulfing floods of wanton misuse. A large vocabulary is, of course, only a gun belt. We still have to learn to shoot straight and be as quick on the draw as Mr. Henry Fonda. The Greeks were taught rhetoric almost as a matter of course. Every Athenian hoped that his turn would come to sway a group of spellbound listeners. The crowds are bigger now and the speeches more numerous, but they are lacking in ornament and are emphatic only with shouting and repetitious abuse. Perhaps speechmaking should be put back in the school syllabus. One thing it would teach us would be the actor Mr. Alan Dobie's first law. The effectiveness of a line of dialogue, he tells us, often depends merely on the order in which the words are placed. It isn't difficult to demonstrate the truth of this.

In a recent television interview the last of the Gaiety Girls was asked if life in the theater during the early years of this century was really as wicked as we hope. The lady explained that at the time when she was on the stage a peculiar morality prevailed. She described the situation in some such words as these. "We were allowed to accept gifts of flowers, candies, jewels, furs, yachts, castles—but never money."

One's delight in this reply depends entirely upon the list of acceptable presents being arranged in a crescendo of financial value before the actual word "money" (delivered with an abrupt return to a practical tone of voice) was thrown in.

Had the lady said that, though girls were forbidden to receive money from strange men, they might accept gifts

of the most sensational value, the information imparted
would have been exactly the same. The entertainment
content would have been nil. Put as it was, I would call the
observation witty.

Wit is the voice of style.

One might almost define an aphorism as an ugly truth
gracefully phrased, or say that wit is any comment upon
the human condition made in a way that is memorable.
Brevity is not the soul of wit. Truth is its soul and brevity
is its body, but, since by now all truths are forgone, it can
only be the form that we give them that is our individual
contribution. Those who sink to mere trading in facts
insult their hearers. We can be offered only two kinds of
information—what we already know, which is boring,
and what we do not, which is humiliating.

In all that I have written on the subject of style. I have
tried as seldom as possible to mention thought. I would
faint dead away if anyone were to receive from this book
the impression that an effective personality can belong
only to the knowledgeable, the scholarly, or even the
wise, though a little wisdom can help. M. Pompidou, the
former president of France, has said that, besides being a
means of expression, the French language helps to shape
the thoughts it expresses. We could go further and say
this of all language. It is not ideas that bring words to our
lips, but new words that suggest to us ideas we didn't
know we had.

In the days when human relationships were still all the
rage, it was seldom passionate suitors who were eloquent;
the very urgency of their desire made them tongue-tied.

It was young men who read a lot of second-rate poetry who were forever flinging themselves down on one knee before some woman or other. In order to release the mellifluous phrases floating around in their skulls they fell in love and, if they didn't watch out, were loved in return.

In the beginning was the word.

If then so little brain power is required—if all you need to be invited everywhere is a toothbrush and an overnight bag full of words—if the cost is so little and the reward so great—why is not everyone at least a part-time wit?

Alas, the answer is more or less as before. Even more than people mistrust words by themselves, they fear them when they are arranged in a conscious pattern. To speak for a moment in terms of the visual arts, most men are "action" speakers. They throw a few words over one shoulder and then look round to see if by a fluke they have made a sentence. As Mr. Beverley Nichols remarked, if you praise an Englishman for saying something witty, he looks as though you were telling him he had stepped in something.

People dislike wit partly because they hate talking for talking's sake but mostly because they think of wit as a way in which clever people score off those who have minds less nimble than their own.

There are, it cannot be denied, immortal examples of this.

In the upper regions of Parisian society there once dwelt a certain Mme. Strauss. When a concert of extremely modern music was being given, she felt the occasion was sufficiently important to warrant her presence. Not intending for a moment to be aesthetically out of her depth, she invited the highbrow music critic of the day to escort her. At some time during the evening, the man, to display his musical sensitivity, uttered the opinion, "This music is altogether too octagonal." Mme. Strauss remained unruffled. "I was just going to say the same thing," she said.

The hushed, elegant bitchiness of remarks like this is what gives wit a bad name. It leaves not only the person at whom it is aimed but everyone within earshot not quite knowing whether anyone has been insulted or not. Most people prefer what they consider the broad swipes of humor, but even here they are not on completely safe ground. A mere smile can be double-edged.

I was once in a "housewife's choice" sculpture class when the husband of one of the students paid us a visit. As soon as he appeared, his wife left her graven image of me and ran anxiously toward him. From the doorway behind me came this fragment of dialogue:

Wife: How did it go?

Husband: I told them they were all talking rubbish.

Wife: Oh, darling. I hope you didn't actually use the word "rubbish."

Husband: I most certainly did.

Wife: Well, when you said it, did you remember to *smile?*

The only kind of merriment that the man in the street does not fear at all is a joke—a prefabricated anecdote that follows the question, "Have you heard this one?" He likes this kind of humor because such tales are always about two other people, never about the speaker and the listener. This is precisely what makes them styleless. The stories that come out of a salesman's notebook exist in isolation. They are inorganic and almost incapable of taking on the personality of the speaker. Also, though some of them may provoke laughter, all of them provoke other similar tales. An evening spent swapping funny stories quickly takes on the horrible boredom of an amateur variety program. Every few minutes the entertainment stops and then starts again. For a social occasion to have something like the quality of a play, its wit must derive from the very nature of the situation in hand—must be provoked by the diversity of personalities involved.

The only thing that justifies the existence of these tales that people hand round like cigarettes is that the best of them embody a general principle of humor which we can study and put to our own use. We already know that what matters is the order in which we place our words, but this is not enough. We must formulate infallible rules about this order so that, if we wish, we can make our simplest utterances funny.

Whenever people ask us to display or preserve a sense of humor, they want us to accept with equanimity some calculated insult. This request is very annoying but at least it is based on a genuine idea of what humor is. It is detachment. A true humorist is so totally unengaged that he can relax in any situation and evaluate every crisis—

even one brought about by his own folly—from at least two (possibly opposed) points of view. This may seem to be a lot to expect of human nature, but in fact, in some people detachment seems to have known no limits. When Mr. Kierkegaard fell half fainting to the floor at the feet of his friends, they tried to lift up his body and carry him to bed. "Oh, leave it," he murmured. "The maid will sweep it up in the morning."

From a standpoint as remote as this it would be possible to regard any incident as a kind of play, and this is what the ideal raconteur does—using, naturally, the dramatist's two chief weapons, suspense and surprise.

We can observe this theory in practice by examining any well-known joke. Let us take the tale of the faith healer and the child. I choose this one not because it is the funniest story in the world but because its form is classic.

A child goes to a faith healer and says, "Please, sir. Father's ill." The old man explains that these words contain the very root of the patient's malady. "What you mean," he says, "is that your father *thinks* he's ill. Go home and tell him this and in a week's time come back and report on the change in his condition that you will undoubtedly observe." So the child departs, does as he is told, and returns on the appointed day. "Please, sir," he says. "Father thinks he's dead."

Here, almost naked, are the principles on which humor works. First, the words are arranged in such an order that the one containing the surprise is at the end; second, the tragic statement in the second half of the tale is made in the optimistic phraseology set up in the first half. This is all there is to it. I deny utterly that the humor of the story depends upon making fun of the doctrines of faith healing. The laughter lodges innocently in the space between form and content.

In most narrative jokes the denouement is the opposite of what the hearer has been led to expect, but in the very best stories of all, suspense and surprise are in alignment. The audience is given what it feared or hoped and the surprise lies in the excessive degree of fulfillment. In the movie *It's a Mad, Mad, Mad, Mad World* there is a fist fight in a barn. The two men involved lurch and stagger toward the window. The audience knows that at any moment the glass will be shattered, but when finally one man is knocked from one side of the barn to the other, he pushes out not merely the window but the entire wall. This climax is made far more effective than it might otherwise have been by the fact that the audience has been lulled by the sight of the window into imagining that it already knows what will happen.

So far so simple, but the work of humorists grows harder with every passing year. There was a time when every large subject had its own built-in quality. Death was solemn; sex was naughty; virtue was sacred. The iconoclasm of our age has altered all this. If we look at the work of humorous authors who were writing before the First World War we notice this at once. Readers of the short stories of Saki experienced a delicious sense of outrage at the flippancy of Mr. Clovis Sangrail almost every time he opened his mouth.

"When I was young," said Clovis, "my mother taught me the difference between good and evil—only I've forgotten it."

"You've forgotten the difference between good and evil?" gasped the princess.

"Well, she taught me three ways of cooking lobster. You can't remember everything."

Nowadays, when moral distinctions have been erased from the mind not only of Mr. Sangrail but of a whole generation, no one coming across the passage I have quoted (possibly inaccurately) would know at what he was being asked to laugh.

A modern humorist has to impose the required emotional quality on his subject matter by his choice of words and tone of voice before he can start to debunk it. While beginning to tell his story in one mood, he must, like a good thriller writer, foresee the contrary ending.

Though we have said much about vocabulary, it must not be forgotten that words are yet another area in which freedom may bring havoc. The novelist Lord Snow says that every man may be permitted his own clichés. But this is not enough. He must invent his own clichés and then persist in them until, if other people use them, their source is instantly recognizable. Thus plagiarism shall be turned into homage.

Once we know the words we are halfway to ruling the world—or at least the society in which we move; we become masters of ourselves because words are the salve with which we heal the wounds inflicted on us by our actions; we can take command of others because to talk about people is to diminish them; and we can control the future because the way to bring events to pass is to predict them.

THE PROFESSIONS
AND STYLE

It is, of course, not enough merely to make sure that the foundations of your home life are solid. You must then decide what you are going to do in the outer world. Some of my readers may be so old-fashioned that they still have jobs. If this is so, they should make every effort not to take work which involves them only with things. These might be called the "making" professions; they should aim to find employment that brings them perpetually into contact with people. They will then be able, during every waking, working hour, to polish their techniques of self-presentation. Work of this nature can be described as a "doing" profession—only one step away from the Profession of Being to which all true stylists aspire.

There was a time when quail traveled by boat from the place of their birth to the hotels where their public awaited them. The journey was long and slow. So that restaurateurs might avoid offering their customers a delicacy that had grown old and gray in transit, the birds set out on their travels alive and in early chickhood. They were placed in the care of a man whose duty it was, apart from seeing to their general welfare, to feed them in a manner to which nature had accustomed them. Between tournaments of deck quoits, his time was spent taking handfuls of grain into his mouth, chewing them until they were a soft, warm, moist pulp and then disgorging them into the beaks of his little friends.

It would be difficult to categorize this profession in any way, but from the viewpoint of this book it could safely be said that it was a function hard to fulfill with very much style.

There are many occupations that could be defined in this way, and stylists ought, if possible, not to undertake any of them. Never do for a living any job in which you cannot add what you are to what you do. It is a mistake to think of style as something that a man puts on like evening clothes only after office hours. He should discover or invent an everyday occupation onto which he can weave the tapestry of his life-style. This is less of a problem to young people than to the rest of us. They, when they have used up the last available grant for the last possible subject, can go on public assistance and pursue self-realization without financial worries. Older people are often ashamed to do this. They and such students as desire a higher standard of loafing than the government can offer must look for some way of combining style with profit.

THE MAKING PROFESSIONS

There used to be ways of earning a living in connection with which the word "style" was mentioned a great deal but in a manner that was misleading. These were the crafts, mere mention of which brings on in some people a fit of nostalgia. They like to think of a young man being apprenticed for years to his elders and betters, then blossoming out on his own and finally being able to produce artifacts which amaze the layman with their high gloss.

All such occupations should be avoided like the plague. They are the making professions. They include manufacture, the visual arts (writing is of course one of these), and all manual labor. When the word "style" is used in this context it refers not to the practitioner but to the thing he creates. Whether he makes a cake or a cabinet or a castle, he is in danger of being upstaged by his creation. Occasionally he may even be made ridiculous by it. Cries of dismay have been known to spring from the lips of readers of lurid romances or muscle-bound adventure stories when they first catch sight of the ill-favored authors of these books.

The fact that the making professions provide such a hopelessly indirect route to a life-style is what makes it seem so foolish of the government to offer students seductive grants for the visual arts.

By conspiracy aesthetic standards have been lowered until art has become a game that any number can play. Even so not all students are deceived. After a year or two

many of them become aware that in entering an art
school they have walked into a blind alley. According to a
survey conducted by the National Union of Students,
only one graduate in three can get a job as a direct result
of his training. Even for the highly gifted the prospect is
bleak. The rest find themselves living in open sin with
their lack of talent as with a demanding but hideous
mistress. What appears to be their laziness is actually
their despair.

It might be truer than describing art schools as culs-de-
sac to say they are halls of mirrors. With every movement
a pupil makes his image proliferates. All these colleges
can safely promise is to teach people to teach people to
teach people. When you realize that many students have
no more talent for teaching than they have for art, you
see how lamentable the position is. We are becoming a
nation of art teachers—reluctant art teachers—and this
has come about at a time when it is universally agreed
that art cannot be taught.

In vast colleges so weirdly designed that birds can fly in
at the windows even when they are shut, and so hastily
built that almost at once fissures appear in the walls wide
enough for you to insert your fingers, members of the
staff wander about like expatriates in a refugee camp.
Not even wages of more than £2 per hour can persuade
these men to teach what in a sense no longer exists.
Called by their first names and disguised as teenagers to
help them shrug off the burden of their status, they drift
along the cracked corridors praying that the cup of
pedagogy may pass from them or, while their charges

neck or fight in other parts of the building, they sit in the staff room, smooth out their pay vouchers with ivory hands, and sigh.

Even if, by a miracle, an art student does become a professional painter and succeeds financially, he inherits a styleless tradition. An artist spends his days adrift on a sea of turpentine or stranded in a desert of plaster, while his treasured creation is taken from this place to a place of execution where it is hanged until it is dead.

Artists themselves are aware of this predicament. The notices of Mr. David Hockney's exhibition at the Lane Gallery in Bradford ended with these words: "Hockney himself will be there talking on Wednesdays at 2:30 P.M." I do not think he arranged to do this because he did not trust his pictures to speak for themselves. I believe he did it for his own sake—because embracing one's devotees by proxy is unsatisfying.

In England, if anyone is shown a huge great piece of concrete with a hole in the middle of it, he at once says, "It's a Henry Moore!" but if he saw Mr. Moore himself, it is probable that he would not have the faintest idea who he was. All that chiseling, all that chipping has therefore been in vain.

Other artists whose life-style overflowed the banks of their art dealt with the problem in other ways. Mr. James McNeill Whistler talked—and not only on Wednesdays. He went into a perpetual double act with Mr. Oscar Wilde almost every word of which seems miraculously to have been recorded. Mr. Augustus John's life-style was an even greater triumph. He did not bother to be witty;

he merely lived in such a way as to push the current image of a painter to its limits. He was the prototype of the Chelsea Bohemian of his day—bold, bawdy, bearded. The strange thing is that, now that the red glow of his personality has been extinguished, his pictures exhibit a faded charm—almost the opposite quality to that which characterized his life.

Another successful stylist is Mr. Salvador Dali. Being a surrealist, he started off with special advantages. Surrealism was the last artistic movement to draw any concerted response from the British public. The word "surrealism" passed instantly, like vaccine, into our blood stream and can never be dislodged. Mr. Dali is not generally considered to be surrealism's greatest artist but he has certainly been its busiest public relations officer. His antics were in all the papers; his photographs, wearing his toasting-fork moustache, were in all the magazines; and in case any corner of his personality should remain unlit by the spotlight of publicity, he wrote the story of his life in which almost everything that other people would regard as frightening or disgusting was described as beautiful.

Compared with Mr. John, Mr. Dali could be seen to be working very hard at his image. This caused him to be accused of "showing off." Rightly he ignored this. No stylist should ever be afraid of this line of criticism. Without an element of vulgarity no man can become a work of art.

We must not wonder too greatly at manifestations of childish inconsistency in artists. The visual arts are a regression. Naturally children express themselves in this way; they see such a long time before they speak. But speech, once mastered, is a means of communication so subtle and, more important, so social that to continue to use any other is perverse. To insist in adult life on smearing surfaces with pigment or—an even deadlier giveaway—plunging one's hands into soft clay can only be regarded as infantilism.

If any of the making professions comes under the heading of art, the sticky cliché with which the cracks in it are pasted over is "The Joy of Creation." In art as in life, there may be a few moments of ecstasy in the act of conception (don't count on this), but bringing anything to birth is usually a long, painful, and appallingly styleless process. Moreover, it takes you away from people.

Those forms of the making professions that can only be classed as crafts are said to be graced by "The Dignity of Labor." In fact, of all the burdens placed on the bowed back of humanity the most humiliating is work.

Of the making professions the most civilized is the making of books. It is therefore among authors that we can most easily compile a sizable list of men and women who have transcended their medium and made sure that they themselves displayed more style than their mere creations.

The pleasure to be derived from sweeping statements lies in the saying of them rather than in the hearing. Sir Max Beerbohm's book of collected theatrical criticism is in parts slightly annoying to read but it must have been a joy to write. One among many categorical statements that he makes is that of all temperaments the histrionic and the literary are the furthest apart. It is, of course, possible to define these two types of personality so that they seem directly opposed to one another, but Sir Max clearly means that they are seldom found in the same man. Stage people, it is true, are seldom literary even now, and in Sir Max's day they were often not literate. Writers, on the other hand, are often theatrical in the extreme. (Is burying your poems in the coffin of your true love not histrionic?) In every author there is really an actor who in some cases does get out.

In the days before youth was considered a virtue in itself, a lot of young people felt ill equipped for life at firsthand. They read about the world instead of going out to meet it and they wrote poems, essays, etc., instead of challenging it face to face. If their writings got published but were not well received, they quickly scorned the world before it scorned them any further and wrapped themselves in the literary temperament. If their work was acclaimed, they emerged from their caves and took to the market place and in many instances never wrote again—not even home.

Others continued to be professional authors but combined this activity with more direct forms of communication. One obvious example of this behavior was Mr. Charles Dickens. When his fame was assured, when he knew for certain that he couldn't go wrong, he emerged from between the covers of his books and, looking as much like Mr. Emlyn Williams as he was able, strode into the light of day to read his works to the public. Thereafter he used his writings to adorn himself.

Mr. Mickey Spillane used himself to adorn his writing. When he could no longer bear to see other actors playing the role of his hero, Mike Hammer, he himself took the part in one last wild film, *The Girl Hunters.* Both these devices are directed to the same end—to turn a making into a doing profession.

Elsewhere in the book by Sir Max Beerbohm from which I have already quoted the author says that the gift of conversation is denied to writers. This is not true. Mr. Wilde was a compulsive talker and a successful one. The truth is that people who do not have a ready wit and are unwilling to acquire one had better go home and turn their tear-stained faces toward long-suffering typewriters. There is envy here. Mr. Peter Barnes, author of the play *The Ruling Class,* says that he mistrusts literary parties because he feels that the sparkle that abounds at such functions should rightly be on the written page. How parsimonious, if not actually perverse, this statement is! Are there really not enough epigrams to go around? And if by some awful chance we should run short of them, should life rather than mere literature be declared the disaster area?

Mr. Wilde seems to have thought not. "I have put my talent into writing," he said. "My genius I have saved for living."

These words should be woven into samplers and hung above the beds of all would-be stylists. The values seem so right. It is a pity that Mr. Wilde's subsequent actions belied his words.

At a first glance Mr. Wilde seemed to have a cast-iron life-style—almost to have invented the idea. His appearance was against him. Everything else, however, was in his favor. He had had an education in the arts; he had enough leisure to cultivate his tastes; he had talent and he had fame. In the days when the music halls were still open, a good test of fame was whether a comedian could raise a laugh by mentioning your name. This test Mr. Wilde passed at "A" level. In *Patience* we have a whole opera the humor of which was originally heightened by recognizing that it was Mr. Wilde who was being lampooned. Even at a time when dandyism was a fashion, he was a conspicuous aesthete who walked down Piccadilly with a poppy or a lily in his medieval hand.

Everyone seemed to condone this posture. Because he had as yet done nothing to weaken the structure of their society, audiences laughed at his jokes even though they were about the delights of sin and the boredom of virtue. People found them just shocking enough to be delicious.

Mr. Wilde took his popularity at its face value even though he knew his life could not be judged so superficially. He really seems to have thought that he could rise above common morality while almost certainly knowing that he was already in danger.

When I was young and used to waltz round London's West End, the name of Oscar Wilde was on the lips of every male and female prostitute. He was the subject of innumerable dirty jokes, of diatribes of vilification, and of emotional speeches about sexual freedom. Whatever the moral attitute adopted, the legend on which it was founded was always the same. Mr. Wilde had met a gilded youth called Lord Alfred Douglas, fallen in love with him, and thrown the world away. I believed every word of this.

Now, nearly fifty long dark years later, when I have known so many homosexuals who could not be prevented from telling me the stories of their lives, I think differently. Strange tales have I heard but none describing a conversion to homosexuality of Pauline suddenness. From this I conclude that Mr. Wilde had been at least bisexual for a long time before he met Lord Alfred Douglas.

Even the movies about Mr. Wilde did not try to drag any love into the relationship between him and Lord Alfred, although love is a religion to the film industry. They made it clear that the younger man procured youths for the older and then nagged him for money, stopping short of blackmail only because it wasn't practical. He would have been dragged down with his victim.

There is, of course, nothing wrong with being sordid— it makes a rattling good life-style; but Mr. Wilde floundered between sordidness and an almost fatuous conception of beauty. He festooned the dung heap on which he had placed himself with sonnets as people grow honeysuckle around outdoor privies.

His life-style was already weakening. When he sued the Marquis of Queensberry he withdrew it altogether, as sharply as a chameleon retracts its tongue. He should have known that he could not suddenly invoke the laws of a society that he had so volubly professed to despise.

Who did he think would come to his aid? When you make fun of people, they laugh not because they feel free of your mockery but because they feel helpless. When they are no longer at your mercy, the laughter dies on their foaming lips.

Within the terms of his life-style Mr. Wilde need never have brought any legal action. He could have feigned to be above confession and denial. Of his friends some would have known that he was queer; some would not— would have been impossible to convince. All implored him to go abroad for a time. Mr. Wilde took no one's advice. He stayed because he was a spiteful man and also because he couldn't bear to leave the stage. This was commendable, but if you are going to hog the limelight you must know exactly which of your acts you are going into. Mr. Wilde did not. When he came to court he tried everything by turns. Asked if he had kissed one of the young men mentioned by the prosecution he said he had not because the boy was not pretty enough. I would call this remark pert rather than witty. It would only just have got into one of his plays. Later in the same trial he had the nerve to cite in his own defense poor old Mr. Plato, who died a philosopher and came back as a spinster's alibi. He also launched himself into a long speech holding up the "love that dares not speak its name" as a love that is pure. For all I know such love may exist, but the time to go on about it is not when there has been read out in court a list as long as your arm of boys you never met except in heavily curtained rooms in Oxford.

From the verdict onward Mr. Wilde seems to have

fallen apart still further. In jail he attempted a complete reversal of style, which can be very effective, but it is an unalterable law that the new image must be brighter than the one that has been abandoned. Here this was not the case. The verses that he wrote before his imprisonment could not be described as anything more than pretty, but the "Ballad of Reading Gaol" contains almost every hitherto forgone epithet imaginable.

I have known personally men who have served sentences twice as long as Mr. Wilde's and who immediately afterward have taken up their lives again, neither refusing to mention nor insisting on dilating upon their experience. Why did Mr. Wilde emerge from prison a broken man?

As I see it, it was his style that was broken. It was never a part of him but rather it was a sequined Band-Aid covering a suppurating sore of self-hatred.

Mr. Wilde made fun of his contemporaries by reversing their moral values in a series of dazzling paradoxes, but in reality he was bound hand and foot by their ethos. He believed as firmly as Queen Victoria in good and evil. One cannot help feeling that even homosexuality was attractive to him primarily because it was so wicked. There certainly has to be some secondary explanation for this part of his life. He conducted his affairs with men so very oddly. His attitude was quasinormal. Whoever heard before or since of a man taking boys to private rooms in expensive hotels rather than among a few convenient bushes?

Perhaps we should amplify the well-known utterance

quoted earlier to read, "I have used my talent to satirize a society by which I secretly felt enslaved; my genius I reserved for disguising a personality of which I felt ashamed."

One might describe Mr. Wilde as someone who used a very considerable literary talent as a trellis on which to trail an almost overwhelming personality. At this game he was one day to be beaten by a woman who may have been a little girl in America when he was striding through that continent lecturing the barbarians on aesthetics. This lady's name was Miss Gertrude Stein, and she was greater than Mr. Wilde in that she used a small—some said a nonexistent—gift as a colorless fluid in which to suspend a monstrous *ego*.

It is heartening, when comparing these two so very different writers, to be able to point out that, though both of them sought world-wide admiration, neither was beautiful in a universally acceptable way. Mr. Wilde's appearance was such that on first seeing him Lord Alfred Douglas was sorry for him, while Miss Stein, whether in photographs or in Mr. Picasso's portrait of her (painted in her absence), looked like an obstinate middle-aged man.

Though Miss Stein had no beauty, she did have other things. One of these was enough money to indulge her whims. In accordance with the laws of style, she did not allow these to multiply. By the time the world began to hear of her, she had only two. The first concerned her environment.

She wished to live where culture grows wild. She

therefore left America for Europe. "Art in this century," she said, "is something done by Spaniards in France." Once across the Atlantic, she set about picking artists in handfuls. This she managed by the simple and practical expedient of buying their pictures. Having made the acquaintance of Mr. Picasso, Mr. Picabia and others, she cemented her relationship with them by inviting them to dinner and arranging the table so that each painter sat opposite one of his pictures. All this demonstrates her singleness of purpose and her cunning, but she must also have possessed some other quality. It is well known that the only solid food taken by artists is the flesh of patronesses' hands, which distasteful fare they wash down with swigs of absinthe. Yet no one ever seems to have despised Miss Stein. Those who were not members of her immediate circle wished they were. Almost every American of that era who had any artistic pretensions whatever ran all the way to her house the moment his boots struck the beaches of Normandy.

Miss Stein's second whim concerned herself. She was not content to carry a banner in the great cultural protest march of her decade; she wanted to wield her own pitchfork or ax or scythe at the storming of the establishment. It was not enough to rule her own coterie; she wanted, like Yum-Yum, to rule the earth.

All around her the visual arts were being convulsed by cubism, futurism, and other forms of abstraction. Miss Stein decided to perform a parallel act of liberation for literature.

Her crusade was doomed. Color and, to a lesser extent,

shape—especially on a huge scale—evoke a certain emotional response whether they represent recognizable objects or not. Words have almost no power outside their meanings. Alliteration, onomatopoeia, assonance, and all the other Tennysonian tricks work only in alignment with the sense of the phrase they decorate. If they move in a contrary direction they become inert.

Miss Stein disregarded these eternal laws completely and, to provoke her readers further, did not call her writings verse. Had she done this, like Mr. e.e. cummings, she might somewhere have won a little dazed acceptance. Miss Elizabeth Barrett's maid was not the only person who did not expect to understand poetry. Some would rather die. Ignoring this loophole, Miss Stein referred to her shorter pieces as stories though they had no narrative content whatsoever. Even to say this gives to the uninitiated no hint of the anarchy that rages therein. One of these items ends thus—"Stew stew than."

Miss Stein also penned portraits of people whom she knew, but here again in none of these did there arise any image of the person whose name was superscribed. As an illustration I quote the beginning of "The Portrait of Constance Fletcher": "O the bells that are the same are not stirring and the languid grace is not out of place and the older fur is disappearing . . ."

In this style Miss Stein wrote a great deal. She had to in order to convince the world that, even if she was mad, she was sincere. In this she ultimately succeeded. That her work was printed is not surprising. She paid. After a while she did better than this. Her sheer persistence—

that massive belief in herself that is one of the prerequisites of style—bludgeoned some publishers into taking her seriously. When she was sufficiently famous for *Harper's Bazaar* to print a parody of her prose, she wrote at once to the editor and asked why the magazine did not publish the real thing. "It's much funnier," she pointed out. The contribution she sent appeared in a subsequent issue. From this tiny incident we see that she was prepared not merely to accept but to invite ridicule while never openly admitting that she wrote as she did for laughs. In this she resembled those actresses who, playing in melodrama, if they fail to make their audience weep, overact until they provoke laughter. There is a vertiginous tightrope stretched between pomposity and clowning that all stylists must be prepared to walk.

When asked why she wrote at all, Miss Stein spread out her arms and cried, "For praise; for praise; for praise."

It's a wonder she didn't starve. She would have if she had not been ready to accept, instead of adoration, a few crumbs of blame. Yet she would not gobble up any old crusts of publicity. Along with all other stylists, in the midst of excess she drew the line somewhere. She lived in France—that country to which Lesbianism is what cricket is to England—yet, though she looked like a man and lived for a lifetime in domestic simplicity with Miss Toklas, she seems seldom to have spoken and never to have written about her love life. She wanted literary notoriety or none.

Nothing remains of her empire. Of all her books the most readable is *Everybody's Autobiography*. This is doubt-

less still on the shelves of many public libraries but I
should be surprised if it is often borrowed. A few years
ago, in an effort to revive her memory, someone wrote a
biography of her called *The Third Rose*. It seemed to
cause little stir. The difficulty of recreating the quality of
a stylist is almost insurmountable. Stylists are not interesting for what they write or for what they do but for
something that they are.

Not only without looks but without much talent or
even sense, Miss Stein, by sheer force of personality, got
to know everyone she wanted to know and became a
household word throughout the American-speaking
world.

Among the many Americans who visited the court of
Miss Stein was Mr. Ernest Hemingway. On one occasion
he held a very significant conversation with her. It was
about homosexuality. Miss Stein defended its practice
among women—as well she might—and explained that
female homosexuality was less perverse and cleaner than
homosexuality among men. She evidently thought that
all physical relations between males were sodomitic. Mr.
Hemingway did not argue this point but condemned all
sexual deviation. His views on the subject were so strong
that he said a man must be prepared to kill rather than
submit. Usually, when an overture of this nature is made
to a grown man, his slightest frown is sufficient to cause
the other party to go and join the Foreign Legion. It is
therefore difficult to imagine where, outside the novels
of Mr. David Storey, the rape of one man by another is a
likelihood.

Mr. Hemingway's remark sounds less like an expression of considered moral condemnation than an almost involuntary cry uttered in some obsessive, lurid dream. This perpetual nightmare was what provoked and maintained the author's life-style.

Though not exactly autobiographical, his novels were adventure stories whose heroes were nearly always excessively masculine men of action and, as Mr. Hemingway grew older, so did they. Unlike Mr. Kipling and Mr. Galsworthy, two other writers who liked the idea of a man's world, he did not seem to think of women as a nuisance or as a means of acquiring property. He described his heroines with affection but not with great understanding. To have identified himself with them even for the purpose of literature might have brought back the terrible dream of androgyny. He may even have feared that there was something effete about the occupation of writing fiction of any kind. To atone for this he displayed in his life a degree of courage that came near to making it a parody of his books. He fought in all the available wars, fell out of airplanes, and shot everything that moved throughout the length and breadth of Africa. In the end he even shot himself. People always refer to this last act as "the final tragedy." I cannot imagine why. In a way all death and, because of it, all life is tragic. About life the hero of *A Farewell to Arms* says, "It kills the very good and the very gentle and the very brave impartially. If you are none of these you can be sure it will kill you in the end but there will be no special hurry." Surely it is only in this general sense that Mr. Hemingway's death is tragic.

A man may keep on writing far into the night of senility—many authors do—but if he is a man of action he cannot, in the fullest sense, continue to live those last years. To me Mr. Hemingway's suicide represents a triumph of style over life. It shows that he regarded his existence as a work of art requiring a definite outline. His action puts Mr. Wilde for all his protestations to shame.

Unlike Miss Stein, Mr. Hemingway did not have to perform the feat of balancing his prodigious personality on a slender literary gift. He wrote a great deal and, right from the beginning, his work was almost universally admired. His problem was therefore the opposite of hers—how to stomp about in the flow of his fiction without drowning. He succeeded.

Except for a sense of humor, Mr. Hemingway had everything. He was as handsome as the sun; his constitution seemed almost indestructible; he had talent; he had fame; and after movies had been made of "A Farewell to Arms," "The Snows of Kilimanjaro," "The Killers," and "The Short Happy Life of Francis Macomber," he was rich. What is more, he made the fullest use of every one of these attributes. Of all the writers whom I can call to mind, he is the one whose life-style was not only in keeping with what he wrote but transcended it.

In a sense he could be termed one of the martyrs of style. The price he paid for living as he did must have been enormous. No one is by nature as masculine as all that. But then we should not expect a life-style to be free. All we may ask is that it shall be worth what it costs.

If Mr. Hemingway suffered for the sake of style, it was his secret. The style itself was the opposite of martyrdom. It was not he but his heroes who always died for their ideals; their author triumphed over all enemies and all obstacles. However, martyrdom can be a style. It is for Mr. Norman Mailer.

He is just as rugged as Mr. Hemingway but more modern. Because morality has changed, he has constituted himself not a hero but an antihero; he has become less a man of action than a symbol of subversion. Mr. Hemingway championed the underdog. In the war in Spain he was on the losing side, but we must never say that he fought for communism because he knew it could not win.

It is Mr. Mailer who espouses lost causes. While Mr. Hemingway may have been the victim of neurotic self-assurance, Mr. Mailer is the exponent of boisterous self-doubt. He positively embraces failure and has compelled the world press to snap him in this compromising position.

His marriages are solemnized like wrongs hushed up, but his divorces are reported in as much detail as a boxing match; if he ever rejoices he must do so in a whisper, but his curses are on a coast-to-coast network. If he were to receive an award we should never know about it, but when he is clapped into a dungeon the police can hardly reach him through the convoy of cameramen. Since it is at his special request that the photographers are present, we must say that disgrace is part of his style. Defeat is the medium through which he fully intends to reach out and touch his audience.

After the success of *The Naked and the Dead* Mr. Mailer need never have got out of bed again. If he did so it must be because he did not feel that he had yet entered the profession of being. With *Why Are We in Vietnam?* and *The Armies of the Night* he moved forward from the writing of novels to the more direct profession of pamphleteering. All his later work, however entitled, is really the bulkiest letter ever delivered to the President—any president. The only blemish on this mammoth gesture is the rejection in his writing of the word "I." Many writers have used the fictional first person; Mr. Mailer has invented the autobiographical third. It is an ingenious device but seems a little coy in so bold a man.

Needless to say, the mere writing of pamphlets soon ceased to be enough for Mr. Mailer; he now also acts them out. In this way style has been brought to protest. To crystallize his rebellion, he has produced, directed, and acted in a movie called *Wild 90*. In real life something might go wrong—or what other people would call right. A policeman might for once be merciful to one of his victims. But in a film of his own devising Mr. Mailer is certain of suffering in person and forever the desired injustice. He fixes eternally before our eyes the blackened image of authority against which he shines so brightly.

Mr. Fred Astaire once said to Mr. Jack Lemmon, "You're at a level where you can only afford one mistake. The higher up you go, the more mistakes you're allowed. Right at the top, if you make enough of them, it's considered to be your style."

He must have been thinking of Mr. Mailer.

Mr. Mailer transformed social defiance into an art, but one has come after him the latchet of whose hair shirt he is not worthy to unloose—someone who flouted the conventions and the law for no reason whatsoever other than because it was his destiny.

Mr. Mailer chooses defeat; degradation is the style embraced by M. Jean Genet. He committed so many crimes that finally he was condemned to life imprisonment. The sentence was never carried out because M. Jean Cocteau successfully petitioned the French government for a pardon. Thus was justice tempered with artistic snobbery—almost with idiocy.

If a man's crime is theft, the only extenuating circumstance that can be adduced is overwhelming need. This was never discussed. Instead it was pointed out that M. Genet was a genius. As this is another way of saying that it would have been comparatively easy for him to earn a living in a different way, this argument ought logically not to have secured his release but rather to have aggravated his punishment. What an opportunity was waiting there for a display of official style! The judge might have said, "The court was not previously aware of the prisoner's many accomplishments. In view of these, we see fit to impose the death penalty."

His pardon was granted in accordance with the erroneous but almost universally held idea that culture belongs to France and that for a Frenchman to exhibit talent is a kind of patriotism. M. Cocteau is now dead. We can never know but only hope that he was shamelessly exploiting the chauvinism of his countrymen in order to ameliorate the lot of a man who was, in some respects, a kindred spirit. If, on the contrary, he sincerely subscribed to the notion that there is a close connection between aesthetic and moral values, it is a great pity. This kind of confused thinking (which we also see in operation when books are exempted from censorship on the ground that they are well written), leads to words like good, moral, true, beautiful all acquiring the same meaning.

Forgiven by M. Cocteau, M. Genet was then canonized by M. Jean Paul Sartre, who called him the saint of existentialism, and, in a way, he is. If existentialism is the philosophy that decrees that you can only exercise your free will by swimming with the tide but faster, then it is the creed of style and M. Genet is certainly a stylist. Fate bestowed upon him certain special advantages. He was illegitimate and spent his childhood in an orphanage. From there he graduated to a reformatory. If at this point he had not believed in the importance of a life-style, he might have tried to throw off the burdens of his early years, to turn again and become mayor of Paris. He wisely saw that this would be to yield his unique advantage and lose momentum on his journey toward self-realization. Having started downhill, he soon broke into a run toward his chosen goal of degradation. Though his homosexuality seems a thousand times more natural than Mr. Wilde's, part of its appeal does appear to be the opportunity it gives him to heap social opprobrium upon himself.

M. Sartre says that M. Genet knows that he will never commit murder. If this is true, it is to be regretted. The act that would crown M. Genet's life would be to kill someone—ideally M. Sartre—by sexual assault.

Compared with the life that M. Genet has led, writing must seem a very tame occupation, but in spite of this he has found time to write three plays and several books. One of these, *Our Lady of the Flowers*, is described by the author as a prolonged fantasy with which he whiled away the hours while he was in prison. This is explanation enough. While he could not fully live, he wrote. Another of his works is called *The Thief's Journal*. This book is said to be autobiographical, but the reader cannot help feeling that here also we have a large element of daydream. It seems so unlikely that the author, picking his way with whatever care, would never have met anyone of normal sexual dimensions.

In M. Genet's life, the fantasy, the fiction, the reality are mixed with a marvelous degree of homogenization. No one can separate the various elements in his life-style. He is famous because he has written books that have been widely read and his work is widely read because the public knows that he has lived it.

This situation has been made possible by the moral laxity at present fashionable. The forties and fifties were the age of recorded suffering. In accordance with that divine law by which everything always gets worse, the sixties and seventies are the age of recorded depravity. At one time few people read books of a documentary nature. They preferred fiction. They felt sure it would contain descriptions of goings-on that decorum and fear of the laws of libel would automatically expunge from works founded on fact. The odds are now reversed. Biographies and autobiographies contain material as salacious as can be found in any novel.

In this small sense freedom can be an aid to style, but only in the hands of men of strong immoral fiber.

In laying the names of these authors yet once more before the public I am not suggesting for a moment that anyone should read their books or attempt to imitate their prose style. Whether they were good or bad writers is neither here nor there. I am trying to show that, though for some people writing remains forever merely a making profession, for others it can be converted into the starting point for a life-style.

The writing of books can become a very bad habit but to write one is salutary. Everyone should keep a diary not only of events but also of opinions, ideas, dreams—a record of the soul. In Mr. Dali's book there is a section called "Childhood Memories." It contains some lurid, even hair-raising incidents. The reader becomes calmer when he arrives at the next chapter. This is headed "True Childhood Memories."

A diary is useful not only because it could form the basis for an autobiography but also because it helps to clarify thought, brings to light what the writer's character is like, and gives him some idea of how it will seem to others. In this way it becomes a preparation for living out in the open. The tragedy of Mr. Wilde's life is obvious. No one should move in the glare of the searchlights until he has considered all the consequences. When all possible eventualities have been envisaged, the thing to do is to embrace the worst of them. There is nothing like a prison sentence for launching style. The careers of Mr. Wildblood and Mr. Frank Norman testify to that.

THE DOING PROFESSIONS

A list of the doing professions would include teaching, politics, sports, crime, and show business. What all these careers have in common is that, whether the actual wages are received at the end of each week or each month or never, the true reward is experienced simultaneously with doing the work. Also any style they may generate does not pass into an object but remains entirely in the practitioner himself. For this reason most of them involve the rearrangement not of materials but of people. This is done largely by the use of persuasive speech, but style in oratory must not be identified with life-style; the former is merely an adjunct of the latter. Some of the doing professions, such as sport and crime, could be engaged in by deaf mutes.

Competence in any of these professions is beside the point, except that with no proficiency at all one is unlikely to have any of these jobs for long enough to impose one's personality upon it. If, however, a barrister were to win great renown in spite of the fact that all his clients went from the Old Bailey to prison as though on a conveyor belt, or a teacher were regularly to receive promotion though not one of his pupils ever passed an examination, then we should have proof positive of the existence of a superlative life-style.

Compared with politics and show business, teaching seems to have a much smaller and less responsive audience. In the past I think that many schoolmasters felt this acutely. But nowadays teaching is a very stylish profession, because a teacher is nearly always dealing with children and teenagers and they have built in publicity value. It doesn't matter whether his technique is to encourage or to repress them, he can count on the attention of the world.

When I was at my last school I thought that all the teachers were mad. Now that I am as old as the oldest was then, I know that they were mad. They had nothing but their madness to keep them sane. They were lifers in a kind of educational jail. The boys were in for four years (with no remission for good behavior) but we accepted our fate with comparative ease because in those days the whole world was a kind of open prison for the young. The masters were grownups and therefore could have known happiness. It must have irked them almost unbearably to have to spend fourteen weeks at a stretch cut off from everything but one another and the horrible, horrible boys. Very few of the teachers owned cars. Perhaps their salaries did not permit them. If they wanted to catch a glimpse of the outer world, they had to summon up the energy to walk to Uttoxeter, take a train to Derby, and wait there for another train to London.

Apparently this problem of how to keep one foot in reality and the other in education was solved very neatly by Sir Charles Stanford. He gave his lectures on musicology in the waiting room of Cambridge station and made sure that they lasted exactly as long as it took the train he had just left to go on to Ely and return.

Derbyshire is abysmally much further from London than Cambridge, so the masters who taught me could not do this. They had nothing to occupy their spare time but the refinement of their idiosyncracies.

One teacher, who was undoubtedly far beyond the age of retirement and must therefore be presumed to have liked the life he led, used to conclude the last history lesson of each term by leaving the classroom while still speaking. We heard only half of his final sentence before he disappeared into the cloisters. For the second half we had to wait until the beginning of next term, when he entered the room on the first day concluding what he had begun to say eight weeks earlier. He said this retained the continuity of our education in spite of the unpleasant interruption of the holidays.

The difficulty for these men was that none of them could extend his personality beyond the limited and reluctant audience of his pupils, who quickly knew every gesture. All this is now changed. The slightest incident that takes place in a school is instantly written up in the papers and the participants photographed for television. Masters who wish to flex the muscles of their life-style have only to take up an attitude that defies the board of governors and they can remain in the public eye for weeks.

Among the men who have done this are Dr. Martin Cole, who taught biology to his class by showing them a Soho-type movie with an educational sound track; Mr. Searle (not, alas, the creator of St.Trinian's), who published poems written by his pupils although they satirized not only the school but the whole neighborhood in which it was situated; and we have Dr. Craig, who taught literature with one hand and was accused of spreading communism with the other.

These are the hemlock boys. They carry on the tradition that began with Mr. Socrates, who was, I would say, the greatest stylist of them all.

It is possible to extend one's style through teaching without defying the Ministry of Education in any way. For example, the great contemporary teaching stylist was Mr. E. R. Braithwaite. Some years ago he wrote a book called *To Sir with Love*. It caused a great sensation when it was first published, but in that book there was not one sentence about education—about whether geography is easier to teach than history. It was full of what the children, their parents, and the other members of the staff thought about the author. Whatever anyone says he is teaching, his subject is really always himself. Teaching is not for students; it is a way in which the teacher may deploy his life style.

Professor Richard Hoggart is a teacher of great renown. Not only did he formerly rule the English Department of Birmingham University but, when he was Deputy Director General of UNESCO, he was accustomed to ruling Europe. He writes books and, because he gives lectures that are televised, the whole of England could be said to be his lecture hall. Thus he has almost limitless scope for style yet none appears. When he is about to lecture, he makes his arrival at the spot from which he will speak as swift as possible, almost taking his audience by surprise and forestalling acclaim. Nevertheless applause breaks out. With his Jean Gabin stance and his sorrowful beauty as of a betrayed trade-union leader he could keep the clapping alive for minutes. Professor Hoggart stands stock-still and waits for the noise to subside, as though he were waiting for a supersonic airplane to pass out of earshot. He speaks clearly but without any rhetorical devices at all. His subject is the involuntary gestures of what used to be the working classes. He has watched their hands; he has counted the wrinkles on their faces. We prepare to weep, but before we can summon forth our first tear, he stops speaking. His lectures have what the poet Miss Laura Riding would call a Trojan ending. There are no peroratory phrases.

What makes all this especially strange is that one of the professor's favorite subjects is the way people behave in public. He has a lot to say about how people move their eyes and their hands. If he feels that other public speakers express their personalities in the wrong way, where is the right way—the Hoggart way? Should we conclude that, though he styles others, himself he cannot style?

The greatest political stylist the world has ever known was Mrs. Eva Peron. She ruled her kingdom in a very personal but at the same time pervasive way. In England spelling primers begin with the words "The cat sat on the mat." No wonder literacy is at a low ebb when the first glimpse of it is this banal and even distasteful piece of information. In Argentina spelling books started with "I love Evita." These were the first words her subjects learned to read.

The crowning moment of Mrs. Peron's entire career was when she rose in her box in the opera house in Buenos Aires to make a speech. She lifted her hands to the crowd, and as she did so, with a sound like railway coaches in a siding, the diamond bracelets slid from her wrists to her armpits. When the expensive clatter had died away, her speech began, "We, the shirtless . . ."

After Mrs. Peron's death her worshipers asked the Pope to declare her a saint. His Holiness refused. Had he consented what a triumph for style that would have been—ankle-strap sandals, a double fox stole, and eternal life!

A stylist's ambition is initially to rule himself and then, charged with this inner certitude, to rule the world or the country, or at least the borough. As soon as we define the pleasures of style in these terms we see that politics is, if not the easiest, certainly the most direct way of achieving satisfaction.

It is no longer likely that a British politician will greatly influence world affairs, but in the first half of this century it was possible. During that time, as the historian Mr. A.J.P. Taylor, himself a considerable stylist, tells us, the most dynamic figure in English politics was Mr. David Lloyd George. For this reason it is interesting to appraise his life-style.

Whatever else he does, there are two things a politician must do if he is to create an effect. First he must find a way of remaining in Parliament long enough to get used to power and for his countrymen to learn to recognize his image. This he does by calling on the loyalty of his friends in such a way that they feel they are espousing a just cause rather than protecting a person and by reminding his enemies of their weaknesses so that they get the impression they are being morally rebuked rather than out-smarted. When a politician has mastered this lesson he must take up a succession of causes which have in common some factor that expresses his life-style.

From the very beginning of his career Mr. Lloyd George decided to champion the underdog. This political posture suited him perfectly. Physically he was small; socially he was totally a self-made man (even his double-barreled surname was his own invention); he really was a boy brought up in an unknown village called Llanystumdwy who grew up to be Prime Minister of England; and he was Welsh. His ancestry was a great help, for Wales is a vast kennel of underdogs forever gnawing at bones of contention and yapping incomprehensibly at their masters' heels.

When Mr. Lloyd George became premier in 1916, his

first job was to defeat the Germans. By then the First
World War had become a free-for-all, but it had started
out as Operation Underdog because England had disguised the need to defend itself as a noble dash to the
defense of Belgium.

Mr. Lloyd George threw himself into the pursuit of
victory with ferocious energy. He chastized the munition
workers with the valor of his tongue, accusing them of
drinking when they should be working, and he shamed
the rich into accepting heavy increases in taxation. To do
this he used the formidable weapon of his declamatory
prose. "We shall march through terror to triumph," he
said. He meant that we should crawl through hardship
but this he did not say because the words, "hardship" and
"triumph" do not begin with the same letter. He also said
that war was a time when men part with anything to avoid
evils impending on the country they love. He meant that
we part with our possessions to save our lives. When
peace broke out Mr. Lloyd George turned his unflagging
enthusiasm to a series of gestures on behalf of the little
people that in sum created the Welfare State. Finally,
many years later, he even championed the Abyssinian
underdog.

Political stylists are usually conservatives. This is not
because tories so often come from the upper classes and
have learned to speak ever so nice but because conserva-
tism is in its nature on the side of style; its object is to erect
barriers—to preserve forms and rituals. The Liberal and
Labour parties are without style; their very names con-
demn them, but even called by other names they would

smell as sour. Their aim is to draw us all toward uniformity.

Mr. Lloyd George was billed as a Liberal, but, as he himself noticed, the liberal intelligentsia hated him while the aristocracy were loyal. On a more personal level, his friends were all conservatives—men like Lord Beaverbrook and Sir Winston Churchill. It was his style they loved. In politics, of course, friendship is a jesting word. Though Lord Beaverbrook helped to put Mr. Lloyd George into power, he later put just as much energy and money into replacing him with Mr. Bonar Law. With Sir Winston Churchill Mr. Lloyd George's relationship was even more ambiguous. The two men admired each other's strength, but this did not prevent them from sending each other up with every breath they drew. When writing his memoirs Mr. Lloyd George could hardly refrain from saying that Sir Winston Churchill's greatness as a politician was flawed by his lust for personal glory. On the other hand, when Mr. Lloyd George made his Abyssinian speech, Sir Winston Churchill remarked to someone that they had just heard one of the greatest parliamentary performances of all time. The words were perfectly chosen; they constituted the highest praise while carrying secret charges of theatricality.

Mr. Tom Jones, the trade-union leader, said of Mr. Lloyd George, "He could charm a bird off a branch but was himself always unmoved." This is almost a definition of a stylist. It is the quality on which Mr. Shakespeare lavished so much praise in the ninety-fourth sonnet:

They that have power to hurt and will do none,
That do not do the thing they most do show,
Who, moving others, are themselves as stone . . .

If we read the second line to mean "who allow others to
think an emotion is felt when it is not" we have a glimpse
of another curious aspect of Mr. Lloyd George's character.

Mr. Lloyd George seems not to have wished his secretary, Miss Stevenson, to see him as a man of stone. Her
diaries often describe him as being moved by something
other—at one time by the death of a friend, at another by
a piece of music heard at a concert, and though he was
the merest shell of a family man, throughout his life he
paraded before her his grief at the early death of one of
his daughters.

Mr. Lloyd George is now said to have possessed
charisma. I take this word to mean that he had power to
influence others based neither on their endorsement of
his actions nor on their admiration for his character.
How else can we explain why his illicit liaison with his
secretary, which was known to all his friends and most of
his enemies, was never used against him—not even by
Lady Margot Asquith, of whom it can be said that
bitchiness was her life-style?

One factor that helped to save him from disgrace on
this account was that, while championing the underdog,
he seldom insulted the overdog. He remained respect-
able. On his first visit to France during the war, he
refused to take Miss Stevenson with him. Even with the

land gray with corpses, France retained her style. A trip across the Channel with an unmarried woman would have seemed naughty. He also took care, when he was at home, to present a united front with his family, although his children did not bother to communicate with him even on his birthday and his wife never made an effort to entertain his political friends with any grace whatsoever.

In this respect his life-style may be contrasted with that of Mr. Charles Stewart Parnell, the nineteenth-century Irish leader, who came out on the side of true love. Mr. Parnell seemed to think he could get Queen Victoria to wink at him while he stood before her with one arm around Ireland and the other around Mrs. Kitty O'Shea. He failed. Faced with the inevitable choice, Mr. Parnell adopted the Irish solution and let go of Ireland. We do not know if his style would have carried him through life as a modern Mr. Mark Antony because, in the same year that he married his light of love, he died.

In a similar situation Mr. Lloyd George would have chosen the English solution and forfeited the Kitty. He had both fire and caution, strength and cunning. Yet in spite of all this, something was wrong.

He once said, "I have come to the conclusion that success unnerves one as much as failure." In making this observation he showed himself to be in some strange way a lesser man than Sir Winston Churchill, who was never for a moment troubled by what Lord Tennyson called "craven fears of being great."

After 1922 Mr. Lloyd George did not hold high office again, though he always remained a political being and,

as late as 1935, won seats in Parliament not only for
himself but also for his son and daughter. The thirteen
years between these dates were passed "in the wilder-
ness." The words in quotation marks were not his own.
They were written by his secretary. All the same he must
have allowed if not encouraged her to think of him as an
exile from his true glory. Why? When he left No. 10
Downing Street he was fifty-nine. How could it matter
any longer whether he remained a cabinet minister? His
name had become a household word and his image as
familiar as Big Ben. He knew that henceforth he would
always move among people of great consequence and
that whatever he did or said would be of interest to the
whole world. Yet he just did not seem to be able to make
the simple transition from a life of direct political action
to the profession of being.

We are told that the wildcat has recently reappeared in
the north of England, but the underdog—a species
whose numbers have been diminishing at an alarming
rate for some years—is now extinct throughout the
island. No modern politician can base his style on the
defense of it. Such dogs as are still with us have learned
not only to look after themselves but also to encroach
with ever increasing boldness upon territory that once
belonged to their masters. Mr. Enoch Powell, only stylist
among our contemporary politicians, was quick to realize
this and to constitute himself the champion of the
overdog.

This gambit required far more daring than was ever
displayed by the father of the Welfare State. Mr. Lloyd

George said in resounding language what people wished they thought—an ever popular ploy; Mr. Powell said what people actually think and wish they didn't. For instance, he is supposed to have remarked, "Often, when I am kneeling down in church, I think how much we should thank God, the Holy Ghost, for the gift of capitalism." In fact it wasn't a gift; we worked at it; but we must not cavil in the face of such bravura as this.

Mr. Powell is not merely a conservative but in fact has stationed himself so far to the right that in his heart he has never relinquished his paternalistic hold on India. He seems to feel himself to be an exile from the Kipling setup. It is therefore quite logical for him to base the political expression of his life-style on the immigration problem.

In my youth I knew a girl whose home life seemed, when she spoke of it, to be as horrible as my own. I evidently overdid my sympathy. One day she rebuked me. "I don't hate my mother," she said. "I want her to be happy and a long way off." Some such relationship as this appears to exist between Mr. Powell and the colored races.

Mr. Powell's frankness upon inflammatory subjects is possible only because of his convincing style. He has a great deal going for him. First there is his name. A critic once wrote of Miss Bankhead that she would never have become a star if she had not been christened Tallulah. The fact that Mr. Powell is called Enoch made him unique in my experience and to most people at least a rarity.

The second characteristic that makes Mr. Powell so formidable is his command of words and the dazzling logic which he makes them serve. Like a true artist he puts this capacity for reasoned argument to the utmost test—sometimes applying it to notions that are by no means self-evident truths. Anyone can find a way to present his hearers with obvious facts; only a political stylist can lend inevitability to the specious.

The ultimate means by which Mr. Powell projects his style is by his presence. This is so remarkable that descriptions of his public appearances read as though he is an actor. Professor Hoggart has said of him, "It is the eyes that do it. They are bright blue and very sharp and they glare straight at you . . . Enoch Powell stares right at you for several seconds at a time . . . The hands seem to take on a life of their own, weaving and clutching through the air as if casting a spell. The voice drops almost to a whisper. . . ."

Mr. Powell's style seems invincible, but he has not transcended the medium of politics, or had a television program of his own or, as yet, written his memoirs.

Mr. James Callahan had ample opportunity to become a political stylist. He was tried and found wanting. Mrs. Thatcher now has her chance, but we need not watch Downing Street with breathless interest. Iron hair, class-change speech, and television exhortation are not style.

No, Mrs. Thatcher's style is simply that she is a woman. She doesn't have to do or say anything. The press will aid her in this matter whatever happens. The papers are full of such remarks as "Mrs. Thatcher arrived at the House of Commons looking dramatic in black." No one ever remarked that Mr. Harold Wilson arrived looking commonplace in navy blue.

We said earlier that a person who is preparing his life-style should avoid playing games for their own sake. It is possible, however, for sports to be the medium of one's life-style. Mr. Jack Nicklaus, for example, receives huge sums of money for his golf performances, and is thus exempt from the charge of playing only for enjoyment.

When my mother was telling me about her schooldays, she said, "I never liked mathematics. There was always only one right answer." Style is free from this irksome limitation, and this is especially true in sports. Each sport has its own style. Fencing is operatic; motor racing is daring; rugby (in America, football) is cruel; golf is rich. The only sport totally without style is soccer (which we in England call football). It is for many reasons an unsatisfactory game. There is something faintly idiotic about a sport whose rules forbid a player to use his hands, which are the most adaptable and efficient parts of the human body. We have become reconciled to seeing someone direct the ball with his feet or his head, but the whole ritual is quite unnatural—more farfetched, somehow, than hitting a ball with a stick. In spite of this soccer is one of the most popular games in the world. In recent years its global appeal has been yet further enhanced by television and by the fact that at last the players are given star treatment and star salaries. The game now involves millions of pounds and stirs up violent feelings including the lust to kill. Into this financial and emotional vortex the players are flung at an early age. Because of this and because the game has no style to offer them, some soccer players become disoriented like child movie stars. For a moment it seemed that Mr. George Best was about to become the Miss Judy Garland of the playing field.

The style of boxing is bravura. Pugilists boast in proportion to their weight. Mr. Muhammad Ali learned this lesson the day he was born. With the exception of Miss Gina Lollobrigida, he must be the most boastful public figure who ever walked the earth. "I am so beautiful," he once remarked, "that I should be modeled in gold." Very few of their heroes have been referred to by sports writers as beautiful, but Mr. Ali's own use of the adjective seems to have set the press free of its inhibitions. "His beauty is his armor," wrote one journalist. "What mere mortal would dare to fell this black god?" Evidently self-praise does not silence the plaudits of others. It is also style-enhancing that Mr. Ali's name is his own invention. It makes him more obviously an auto-fact—independent of parentage, almost of history.

The total independence of style from results is particularly noticeable in sport, and if proof of this were needed, it emerged for all the world to see when Mr. Ali first confronted Mr. Frazier. That event will always be the contest that Mr. Ali lost; it will never be referred to as the fight that Mr. Frazier won.

Since that terrible day others have dawned which have

been even more challenging to style. The idol has been, if not actually broken, at least slightly chipped; what was to have been solid gold is now partly silver. Only the wingless could bring themselves to mock. The rest of us wait to hear the giant grind these misfortunes into jingles, aphorisms, limericks, and to see him stride un-daunted even deeper into the public's consciousness for his whole boxing career is in the Muhammad Ali story only a prelude. He once told one of those sports writers who traffic in life-style that boxing was only a means of bringing him to the people. All the anxiety, all the rigors of training, all the suffering in the ring have been endured not to perfect a boxing style but as a foundation for his life-style.

In modern life a stylist does not even need virtue. It is no longer necessary to be an object of public veneration or even affection. He can be the focus of contempt or even downright hatred. Illustrations of this abound in areas even beyond sports, show business, and politics. As a test of whether you are in touch with somebody, being loved can never be a patch on being murdered. That's when someone really has risked his life for you. If you have chosen depravity as the fluid in which you will suspend your monstrous ego, then you have the most wonderful examples before you—or rather behind you, because most of the ones we can best identify are in history.

M. Gilles de Rais murdered at least a hundred and forty boys in a lifetime. Numbers are not style but it's difficult not to be impressed.

He was a nobleman, and the projection of his life-style was made easy for him by the fact that, while he was rich and owned more than one castle, most people in France were so poor that whole families of peasants left their homes to wander about the countryside in search of food much as, according to Comrade Boris Pasternak, the Russian bourgeoisie did during the early days of the Revolution. M. de Rais caused it to be known that he was not only wealthy but also an extremely religious man who maintained a large boys' choir in his castle in Tiffauges. This was a way of luring into his grasp victims of the age that suited his tastes. When the boys arrived hoping to become part of the choir, he murdered them and ravished them while they were dying or, if they died too quickly, when they were dead.

I expect that rape and murder, either separately or mixed together, fill the fantasies of most men and all stylists. They are the supreme acts of ascendancy over others; they yield the only moments when a man is certain beyond all doubt that his message has been received. Of the few who live out these dreams, some preface rape with murder so as to avoid embracing a partner who might criticize their technique.

M. de Rais was a very different manner of man. He occasionally gave select ravishment parties. He would never have done this if he had been physically inadequate. Orgies are for sexual athletes.

After a while not even a pinch of exhibitionism could prevent his desire from outrunning his delight. He took to riding into the countryside and hunting down his little friends. (Here once again we see the strong connection between crime and sport.) In these sorties he was accompanied by a certain M. de Sillé, not so much, I feel, because he needed help as in order to have his prowess in the field admired by his peers.

When he was finally brought to trial, it was for quite another offense, but by this time rumors about his life-style had begun to spread across the land. On a journey through France he had murdered a boy while staying at an inn and this child had parents who noticed his disappearance.

If proof were needed that style engenders style it could be found in accounts of M. de Rais' trial. His confession was so long and so lurid that the Bishop of Nantes ordered the face of the crucifix on the wall behind him to be covered. Moreover, style was not confined to the courtroom. People came from far and wide throughout France to pray for his soul. Some of these people were the parents of the children he had murdered. If this is not style, it is at least gesture on a national scale all brought about by one man.

At the last M. de Rais cried out, "I am redeemable." Into what shade is the whole of Mr. Oscar Wilde's *De Profundis* flung by this single sentence!

The police, who can hardly be expected to approve of style, tell us that with the passage of time criminals become careless—that in the end they slip up. It seems to me more likely that they grow increasingly daring. There have been burglars who fed the housewife's cat and bandits who kissed the ladies in the plundered coach. These are obvious instances of artists extending their style, but almost all crooks from petty thieves to Mr. Jack the Ripper stamp their crimes with their own pattern. In spite of everything they wish their identity to be recognized. In this respect the most satisfactory life of crime was lived by highwaymen like Mr. Dick Turpin, who were known to many and revered by some for years before they were finally caught. If we think of a life of crime as a way of making something of ourselves, then we see at once that detection is not a piece of bad luck but a consummation.

Penology is another sphere of human activity which has suffered the sad homogenizing effect of time. We have now reached a state of affairs in which, if you steal a sufficiently large sum of money or murder a small enough person, you are given the same sentence. In Mr. Turpin's day the crowning moment of a criminal's life was his public execution. Compared with this, what an anticlimactic end it was to shuffle toward the noose in the presence of a mere handful of hardened, sanctimonious officials! Now murder has had even more of its style wrested from it by the abolition of capital punishment. If this seems a heartless observation it should be remembered that it was not trapeze artists who wanted the use of the safety net made compulsory. It was a group of busybodies who wished to rob these performers of the most effective element of their life-style.

To a religious person, though it disobeys the teaching of the Old Testament, the abolition of hanging accords with the principles of the New. It allows the sinner time to repent—possibly even to atone. This is the most that can be said for it. Life imprisonment robs detective fiction of its formality and murder of its special quality. It is not a merciful sentence though it may seem to be. Mr. Dostoevski has told the world his thoughts as he stood before the executioners. If he had at that moment been offered as an alternative to immediate death the chance to live the rest of his natural life on a ledge a foot wide over a bottomless abyss he would have chosen it. This only means that the body prefers, always tries, to save itself. To the mind, life at any cost is not always preferable. If many free men wish for and some bring about their own death, a man with an eternity of prison ahead of him must think of suicide every day. Of capital punishment it could at least be said that it saved a man from the weakest elements of his own nature.

Crime has in common with sport that it fulfills every man's fantasy of plunging into the midst of terrible danger and surviving. Mr. Sterling Moss once said (and who should have known better than a racing car driver?) that the best way to live is "flat out with every decision mattering"—as though the whole of life were the Grand Prix. In such circumstances as these the stylist's ideal of living in the continuous present throughout his body is not just a vague dream of perfection; it is an immediate and total necessity. On the race track, one must do this or die.

Of all professions the one that offers the most direct route to self-realization is the stage, and nowadays no one need actually go on to the stage. All professions have become also the profession of acting because television has become inextricably woven into our lives.

Stylists must remember that the journey made by an actress from nonentity to stardom is the path that they themselves must tread in one context or another. In the beginning of her career an actress plays a great number of different parts—anything she can get. Later she plays fewer, larger, and more closely connected roles. This is like attacking the problem of style American Indian fashion: the circles of experimentation become smaller and smaller until you arrive at the center of yourself. Once an actress has discovered her ideal role, she never looks forward. She has left the profession of acting and entered the profession of being.

After many long dark years Miss Bette Davis was finally given the part of Regina in *The Little Foxes*. This gave her the chance to say, "Very well, then, die. I'll be waiting for you to die." We could tell from the glaring eyes and the mouth worn upside-down that this was what she had been waiting to say for years. Miss Joan Crawford played in any number of movies in which she rose from rags to riches. Finally in real life she married Mr. Pepsi-Cola, and she had all the luck in the world because he died. This meant that she herself could enter the boardroom of that vast empire. This was a consummation that only the most audacious of her scriptwriters would have dared to include.

Mr. James Cagney says that technique is a matter of letting nothing come between you and the audience. Since we are here concerned less with audiences than with actors, let us rephrase this proposition to read, "In acting nothing need come between an actor and his style."

It is useless to search for confirmation of this statement in the contemporary theater, because there style is not a dirty enough word to be popular. We must look at least as far back as the twenties. This was the last decade before doubt was born and the first in which I myself found that I had views about the theater and art generally.

The man who ruled the theater of the twenties was Sir Noël Coward. His greatest gift was for writing lines that actresses long to say. This insured that his dialogue was always spoken with the utmost gusto. Another attribute expressed in almost all his plays was his invincible snobbery. This quality he shared with a great number of dramatists including Mr. Shakespeare.

From the time of Mr. Euripides to the end of the Elizabethan era, the drama maintained a steady altitude among the nobility; thereafter it began to coast slowly down until it included the middle classes. When Wall Street fell, in 1929, the drama made a bumpy forced landing among the lower orders.

In novels millions of words were written about the poor before 1929—a large number of them by Mr. Dickens, but even he, though he professed to champion the lowly, displayed throughout his work a condescension as monstrous as that showered upon Negroes by

Mrs. Harriet Beecher Stowe in *Uncle Tom's Cabin*. It was
an attitude so universal as to pass almost unnoticed until
the Depression. Then suddenly it became unacceptable.
In the drama, where things tend to fly to extremes, the
poor took on the attributes of gods. Presumably this was
because for the first time in a long while poverty was
being experienced by the people who most often went to
West End and Broadway theaters.

Time and President Roosevelt gradually lifted the
Depression from real life. The drama never recovered.

Novelists could cope fairly easily with writing about the
poor. They could describe in their own glowing prose the
noble sentiments of the out-of-work laborers who were
now their heroes. This is what the wise ones did. The
more foolish attempted to make their characters speak.
The worst offender in this respect was Mr. John Steinbeck. In his books everyone not only thinks the most
high-flown thoughts imaginable about life, love, and so
on, but is allowed to express himself at great length in
similes and metaphors and with heaven knows what
literary flourishes all carefully roughened to give a
sickening plastic semblance of naturalness.

From this difficulty, which Mr. Steinbeck embraced
more or less wantonly, dramatists, because of the very
nature of their medium, could find no escape. The harm
done by the Steinbeck fallacy to writing was unimportant. What was serious was the crippling effect on life in
the theater. Ultimately, many years later, it was to cause
one of the worst sicknesses of our time—the bacon-and-
egg stars.

The playwright who came nearest to discovering a satisfactory solution was Mr. Eugene O'Neill. He made his characters sailors or Irishmen or both. This helps greatly because in Ireland and on the high seas (before conscription) poetic style in speech was a group characteristic.

If dramatists were to abandon the condescending chain of malapropisms that, until the Crash, was the stock speech of their humbler characters and, instead, were to formulate a new clodhoppers' rhetoric, it stood to reason that actors must twist and turn in search of a way to declaim it. Eventually St. Lee Strasberg of the Actors Studio discovered the *method*, but before then the theater went through some very shaky forms of neorealism. Sad to say, so did the actresses.

Stage people find it difficult to distinguish between their private lives and whatever part they are playing in the theater. But what am I saying? They flatly refuse to make the slightest effort to separate them. So, when the theater of certainty prevailed, they were certain; when plays were about dukes and duchesses, they were duchessy. During the twenties, an actress could be seen by all the world to be very different from an ordinary mortal. Not only at the stage door but wherever she went she was—she arranged to be—an object of absorbing interest. She had a face like an egg to which cosmetics had been applied and all her movements were operatic. On taking a seat in a public vehicle she crossed her legs so that one shin was almost horizontal and her instep painfully arched. Then she spread out the ruching on

her dress so that it covered the knees of at least two
gentlemen on either side of her and, in tones like the
sound of Big Ben, remarked, "These trains are absolute
hell."

These antics were to her what a blue apron was to a
butcher.

Every part, including the perpetual one, was played
for all it was worth and frequently for more than it was
worth. When an actress found herself in an underwritten
role, she battled bravely with what she considered poor
material and gave a Delphic quality even to such lines as
"Madam rang?"

At rehearsals the director (then called the producer)
spent all his time trying to prevent the leading lady from
sweetening her performance and, worn out with frustration and fatigue, left the rest of the cast to fend for itself.
As soon as his back was turned the play fell into the shape
dictated by natural selection. The player who was han-
diest with the scythe cut down the others not merely to
size but to the ground. It was this spectacle of a limelit
harvest that made theatergoing a pleasure. The actual
plays were idiotic. This was obviously not merely my
private opinion. It was also the view held by the management. To spare the author all feeling of humiliation, it
was customary to print his name on the program in type
no bigger than that used for the words "No Smoking in
the Auditorium."

After this world of black and white—this theater of
asseveration—came the theater of doubt, a world of dim
gray; the inarticulate spoke or, at least, appeared on the

stage, and actresses, wishing as before to keep their private and professional images all of a piece, relinquished their former occupational style. Even those few who may have regretted having to do this were afraid that, if they went on waltzing in and out of agents' offices with leopards on leads, they might be considered old-fashioned—unable to act the new mumbling, stumbling parts.

Thus another color disappeared from the already diminished rainbow that hovered over metropolitan life.

Before this sad decline began—in that golden age when the world was full of whatever is spare, original, strange—when muffin men rang their bells and lavender sellers sang in the streets—when soldiers walked about London in scarlet coats, when sailors had fly flaps on their trousers and prostitutes strutted to and fro in stockings made of fishnet through which whole haddocks could have escaped—in that halcyon era when high court judges understood nothing and bishops were shocked by everything—when workers worked and servants served, when shop assistants assisted and students studied; in that Homeric time when the weather had not been nationalized and the winters were snowy and summers were sunny and style had never known a day's illness, then entertainment was entertaining—or, at least, that was its aim.

It operated on three clearly defined fronts—the music hall, the theater, and the cinema. To look at television, you would think that the music halls closed in 1910. In fact they flourished well into the forties. Their last great star was Mr. Danny Kaye, who was singer, dancer, mimic, and comedian all in one. Who could follow him? After his appearance at the Palladium, vaudeville fell into a decline. People who, like me, dislike the present and fear the future, look back on almost anything that is dead and gone with misplaced affection. To us the music halls have in retrospect a blowzy charm, but really they didn't begin to shut a moment too soon. A vaudeville or variety evening was excruciatingly embarrassing. We remember the names that topped the bill and forget that seven-eighths of any two hours spent at the Alhambra or the Holborn Empire were given over to watching conjurors, ventriloquists, acrobats, and the uncertain antics of trained animals. These acts extorted from the beholder an unwilling admiration for the perseverance, the concentration, the ingenuity, and sometimes even the courage displayed but never evoked the slightest interest in the artists who were performing. This ought not to have been so, for what a strange assortment of men and women they must have been! We shall never know what makes a man become a conjuror or a juggler because no one will ever have the nerve to ask him why he sacrificed so much for so little. The continuous training required for walking up an unsupported ladder or riding a one-wheel bicycle with a cup and saucer on one's head or controlling the recalcitrant wills of poodles or doves or seals must have made the lives of many music hall artistes as thin-lipped as the existence of a monk but without any specific hope of redemption in the sweet by-and-by. When finally perfected, those skills were placed end to end to make nothing more than an hour-long curtain-raiser for some big name who at last appeared and, seemingly without effort, won all hearts by singing a few songs or making some jokes.

The long-term kings and queens of the music halls were always either comics or singers. Outside of natural catastrophes, the greatest power on earth is personality. If proof of this were needed it could be found in vaudeville. The reason why these forms of entertainment were more popular than any other was simply that, whatever their secret concentration on their work, these men and women never let go of their audience for a moment.

After centuries in which the halls had been ruled either by a woman walking up and down a diamanté staircase and waving an ostrich feather fan or by a little man in a funny hat leaning over the footlights, there appeared someone who, in a sense, was both these people at the same time. If all the great variety artists of England were to appear on one program, the largest letters in the brightest lights would spell out the name Gracie Fields.

Miss Fields sang loud and she sang true. Neither of these abilities would she need today. Plenty of modern singers admit that they are virtually incapable of singing a song as written. Some go further and, starting more or less anywhere, push the note they find themselves singing around until they catch up with their orchestra. They are also not bothered by the question of volume: they use a hand-held microphone. In the twenties, even if this device had been invented, the sight of a girl dressed as though for her first date with the world but yanking an electric cable across the floor as she approached would have evoked pity rather than praise. To see her gripping in her lacquered hand a huge metal penis would have been an affront.

Entertainment in those days was basically the spectacle of someone doing something the beholder could not do. Effort was made to make the performance seem easy for the artiste but never for the audience. An element of wonder now totally absent from entertainment was almost always carefully preserved.

Miss Fields's voice was so marvelously rich that the diva Signora Luisa Tetrazzini (whom Miss Fields occasionally imitated) advised her to train for opera. This was, in the estimation of the prima donna, the highest praise that she could offer. Miss Fields, flattered though she may have been, did not follow the advice. Rightly. As I see it this was an attempt on the part of Signora Tetrazzini to drag Miss Fields down to her level—to rob her of her style. However unusual her voice may have been, it was her personality that was unique.

The underlying message of Miss Fields's image would have little meaning today. It relied for its effect upon class distinctions which time has leveled. The publicity put out for her paraded her simple origins. Whether she ever, in fact, sang in the streets of Rochdale, in her native Lancashire, is beside the point. The legend which I constantly heard upheld when I was young was that she did.

In the days when society still had a class structure,

stardom could manipulate either of two myths. An actress could be rumored to have come from the Balkans or worse, to have been a slave, an orphan, or whatever and to have risen to great wealth and fame. She then, whenever she felt she was being observed, radiated a Christmas-tree eagerness, a gratitude to fate, and all the other attributes of a retired waif. Alternatively she could be presented to the public as a self-propagating divinity who longed to leave Olympus and walk across the wind-swept moors in a filthy trench coat.

Miss Fields combined the essence of both these myths. She seemed to be telling her audience that, though she had once been a mill girl and had now become a highly paid and universally acclaimed singer, she had retained the no-nonsense outlook with which she started. This was her key to the heart of the world—an implied or open mockery of all values based on anything but human worth. This stance she never abandoned.

She wore glamorous clothes but she made fun of them. There was a tiny tradition in the halls that, between numbers, a performer would slightly alter her appearance by taking off a shawl or some other accouterment and flinging it with rehearsed negligence across the piano. Miss Fields, after removing a gold lamé coatee, would take a deep breath apparently to begin her next song but in fact only to say, "To show it's a two-piece. See?" In the same mood she would sing snatches of opera but would turn cartwheels as she did so.

On the evening after she had been to Buckingham Palace for her investiture, she ran her forefinger under

her nose and then said, "Eeh! Mustn't do that now I'm a
C.B.E." When it was known that she was going into a play
called *Interference* with Sir Gerald du Maurier, the press
asked her what it felt like to be going "legit." "Seems
funny, being posh," she said. She did not scorn the honor
conferred by royalty; she did not refuse to co-star with
Sir Gerald. All these things were grist to her Lancashire
mill. In her mockery of poshness there was no protest, no
crusade, no bitterness. Almost certainly she felt none,
but, far more important, such an emotion would not
have been within the ambit of that life-style with which
she seemed to have been born, like Minerva, fully armed.

When Miss Fields was at the height of her fame, a
comedian peppered his act with topical references but
he did not need that convoy of gag men essential to
the safety of modern funny men because the bulk of
his patter consisted of saying the same thing year after
year. If he did not (and one way of varying entertainment was to say the cue but not the pay-off), then
the audience said his lines for him. Often these were not
jokes but the merest catch phrases. This did not matter.
Because his hearers could anticipate his words they
felt they knew the actor and to some extent owned
him. Also, as Mr. V.S. Pritchett, the eminent writer and
critic, has pointed out, to the English the more often
something is said the funnier it becomes. Miss Fields had
a freer method than this. Her songs varied from the
rueful to the raucous, from "Ave Maria" to "In the
Woodshed, She Said She Would," but the nightly reaffirmation of her social attitude gave her image the constan-

cy of a fixed star.

This life-style she projected not merely across the footlights but across the world. When the last of the variety houses closed, she went to rule the island of Capri taking her radiance with her. For tourist trade the Blue Grotto had nothing on Miss Fields.

From the music halls we learn that there is no such thing as being too like ourselves—no such thing as being too predictable. All that has changed since those days is that popularity is now easier to achieve. As we have seen in other doing professions, it is no longer necessary to be lovable.

Before the war the music halls (and England was riddled with them) provided the broadest and most direct line of communication between a performer and his devotees. They were, however, not the only places in which personality could blossom. There were also the legitimate theaters. Plays, in that happier time, were written as "vehicles" for certain leading ladies, so that whatever happened they could make the gestures, say the words, and wear the clothes that in their opinion suited them. Occasionally one of these women would put on a pair of spectacles and her agent would say she was playing a character part, but her admirers were never deceived; they smiled indulgently, like grown-ups watching children strut about in adult clothes. In their natural state people are more interested in people than in anything else in the world, and public taste had not then been sullied by education. Now that commitment has become a fad with both dramatists and theatergoers, perspective has become distorted.

This does not mean that we now have no stars. On the contrary, we have more public idols, drawing more money and having more sequins torn off their backs than ever before in the history of entertainment. For performers who really can sing, all is still well. They can drift happily from television to cabaret to pop festival and back to where they started. It is the others who are in trouble. Because theaters are now given over to ideas and the halls have vanished forever, there is nowhere for them to pretend to act; they are compelled to pretend to sing, for the world has become a clockwork orange. Only sex, violence, and music are now acceptable.

Question: What makes a great actor?
Answer: An unscrupulous agent.

It is easy now to give the right response but it describes a state of affairs that is not altogether satisfactory. An astute agent has a very clear view of the market and this leads him to prescribe a minimum-risk policy for his clients. Thus actresses tend to be modeled from the outside rather than extruded from within.

It was not always thus. In former times a girl who wished to go on stage clung firmly to whatever she considered made her unique. This she added, if she could, to the stock requirements of a leading lady. An actress needs good looks, including large eyes, slightly more than average height, total physical coordination, and, as the theater critic Mr. James Agate pointed out, the nervous constitution of an ox. In 1926 there arrived

in London a young woman who had all these characteristics—some of them to excess. In the portrait that Mr. Augustus John painted of her, each of her eyes is larger than her mouth. She also had two attributes which were her very own—the name Tallulah and a husky voice.

At the time when Miss Bankhead crossed the Atlantic, England was, to most Americans, an almost mythical island where brass toasting forks and copper warming pans grew on trees, but she remained undaunted. To add to the uncertainty surrounding her reception, she not only had no contract but also no invitation. Sir Gerald du Maurier, at that time the dream man of all the world, except to his daughter Daphne, having chosen Miss Bankhead to play opposite him in *The Dancers*, had changed his mind and sent word to this effect. Miss Bankhead arrived nonetheless and changed his mind back again. Even now, when various mechanical devices have narrowed the Atlantic to the width of a babbling brook and when effrontery has become one of the charms of youth, few young women would attempt such a coup. Her boldness paid off.

By the time the play with Sir Gerald du Maurier closed, her strategy for storming the city of London had become clear. Unlike other actresses, she did not wait until a Bankhead part came her way, she began to play in real life the traditional part of an actress with such dedication and such vigor that the plays in which she appeared seemed like a few happy hours of relaxation from more serious business. When accused of this, she defended herself with the words "I take my case to the people."

Such a firm hold did she have upon her destiny that she was able almost totally to disregard the critics. She seems to have done only one thing they asked. At the request of Mr. Agate, she abandoned her Southern drawl and adopted the diction later perfected by Mr. Peter Lawford—the Mid-Atlantic voice. Presumably, Miss Bankhead did this because she could not bear a single one of the terrible things she said to be incomprehensible to her English audience.

Except for a play called *They Knew What They Wanted* (whose title alone was enough to command her respect since it expressed her entire philosophy) her disregard for her material was almost total. A stage manager of my acquaintance tells the story of a rehearsal which had been delayed so long by Miss Bankhead's absence that the producer had given her up, if not for dead, at least for dead drunk. He had just decided to play the part of the leading lady himself when she appeared, her squirrel coat hanging on her elbows, and flopped into a chair. A script was handed to her which she opened by instinct at her first entrance. Her lips moved silently as though in prayer and her forefinger moved along the lines of type like a lie detector attached to the pulse of Mr. Ananias. When, to the consternation of the rest of the cast, she seemed to have turned half the total number of pages in the book, she shut it abruptly. "Well, naturally," she murmured, "I'm not saying any of that." By modern standards she would not have been blamed.

For eight bright years, she acted with a number of leading men (of whom only Mr. Leslie Howard ever ran

neck and neck with her) in a series of plays so artificial that some of their lines now read like a send-up.

Mr. Godfrey Tearle (placing one knee on the couch on which Miss Bankhead is cowering): "And now you're going to pay."

Miss Bankhead: "How shall I pay?"

Mr. Tearle: "In the way that women of your sort always pay."

Of this or some equally frivolous production Mr. Arnold Bennett wrote, "Once again I have seen Miss Bankhead electrify the most idiotic, puerile play by her individual force."

So great a star had Miss Bankhead become by the time Mr. Bennett wrote these words that in whatever piece she was appearing her first entrance was greeted every evening by prolonged applause from what the press had come to call her gallery girls. As soon as this ovation began, she left the action of the drama, stepping over a corpse if the play was a thriller, and came down to the footlights. There she stood bowing and smiling until the shouting and the tumult died. She then said, "Thank you, darlings," and, if this did not set the clapping hands off again, went back to her corpse and screamed.

The real-life part of an actress that Miss Bankhead had chosen for herself she based partly on tradition. In doing this she obeyed one of the primary rules of style. She added her individual style to that of the group to which she belonged. In the public mind music-hall artists are big-hearted, dancers are childlike, opera singers are bad-tempered, and actresses are wicked. From the sin of being on the stage at all, which acted as a springboard, Miss Bankhead dived into hell with an élan which was all her own.

She smoked 120 gaspers a day, swore like a fisherman, drank like a fish, and was promiscuous with men, women, and Etonians. To these vices she added the sin for which there can be no redemption. She allowed—nay, arranged—for all these activities to be known. Theater-goers went to see her on the stage chiefly to marvel that such a debauchee could still speak and still stand.

Her speech and her stance were, as it happened, two of the most individual things about her—at least until she passed from being a public idol to being a universal fashion.

I cannot say if imitation is truly the sincerest form of flattery. It seems foolish to mention flattery and sincerity in the same breath. What we can safely say, however, is that of flattery's many guises, imitation is the most pervasive—the one which the object of veneration can most plainly see.

No one will ever imitate our virtues. It would be too much like hard work. If we wish to have imitators, we must have faults and we must display them. Miss Bankhead put everything on view. Her admirers chose to copy her vocal pitch, which was a mixture of a sigh and an indolent bark, and her posture, which was somewhere between a slouch and total collapse.

To modern minds neither of these things—nor, indeed, anything about Miss Bankhead—would seem very remarkable. It must be remembered that, until her arrival in England, only tragediennes spoke contralto. All other ranks used the eggshell voice now employed only for asking, "Anyone for tennis?" Many ingénues were ladylike to the verge of inaudibility. People like me whose poverty condemned them to the remoter parts of the auditorium knew what was going on on the stage only because of the rigid conventions that formalized the action.

In Act Two of all musical comedies, for instance, after interminable squeaks and groans from the orchestra, the curtain rose on a set riddled with trelliswork and stuffed with paper flowers. Though within the play the time was afternoon on a summer's day, the lights would iris down to a circle two feet in diameter around the face of the leading lady. She would open her parasol and her coral lips would part. Then someone in the gallery would remark, "I bet she's singing."

In the matter of deportment the jackknife posture was universal. A flapper stood with her feet together, her knees braced, and her bottom sticking out, as though she were offering to kiss you across an invisible shop counter.

Miss Bankhead put a stop to this so effectively that I can remember a friend of my sister boasting that she had to have the hems of her dresses taken up at the back because she stood in such a way that, if this were not done, her skirts would dip behind.

By 1932 Miss Bankhead had begun to play her permanent role as a wicked actress so boisterously that she was not merely asked by the management to leave a certain hotel but was expelled by the authorities from the whole country. She went to Hollywood, presumably with the intention of painting the whole world the color she had painted England. She failed. She was put into a number of films with titles like *Branded* and *Tarnished Lady*—the kind of glittering hokum that should have fitted her like an eighteen-hour girdle—but all her movies were made without a single frame of humor and in each she was cast as a victim of fate when everyone knew that, as far as she was concerned, fate was a pushover.

In Britain her fame declined drastically. Once she had been so notorious that if she hailed a taxi and cried "Home, darling," the driver knew where to take her. Her life-style had been so full of invention that for six years hardly a day passed without the papers reporting, if they dared, something she had said or done. Yet when she died, the press had to explain who she was.

If the generation gap means anything at all, it describes the abyss that lies between the people who saw Miss Bankhead on the London stage and them that live in darkness.

The measure of a person's life-style is the distance between the talent and the fame. By this standard Miss Bankhead—indeed, almost everyone—must be placed south of Mme. Bernhardt. That she was called "The Divine Sarah" we may discount. She was so named not because of her skill but because of her nationality. A race of people so grossly materialistic as the French must from time to time utter spiritual sighs. It is a kind of moral balance of payments. Nevertheless, Mme. Bernhardt was at one time a star by comparison with whom most experts found all other tragediennes wanting. After reading the works of Mr. Agate one feels unfit to pass judgment on any subject whatsoever if one never saw Mme. Bernhardt.

Other critics have been less ecstatic. Seeing her in *Hamlet*, Sir Max Beerbohm said that if she had possessed a sense of humor she would never have taken the part in the first place. This is only one man's opinion but evidence can be adduced which cannot be gainsaid. There is in existence a record of her voice. The poor quality of the reproduction and the age of the disk may have lessened the sonority of her speech, but the monotony of her delivery, the lack of range and of rubato are obvious for all to hear. In the entire recitation, the tears never leave her throat.

Even more damning is a movie made of Mme. Bernhardt in the death scene from *Camille*. I have witnessed a screening of this at the National Film Theatre. Awkwardly held at arm's length by Armand as though she were a firecracker (and, in a way, she was), she half-sat, half-lay on a couch wearing a minimum-risk nightdress, clown's make-up, and hair that looked as though she had found it under the bed. Though someone must surely have told her that the picture was silent, the moment the cameras began to turn, her black lips started to gibber and twitch faster than those of a policeman giving corrupt evidence, and, sometimes in unison and sometimes in succession, her arms shot out to punch the air before returning after each sortie to strike her a massive blow in the chest. Then, as suddenly as all this galvanic activity had begun, it stopped. Her head fell forward and one arm swung down beside the couch like the limb of a rag doll.

I am a self-confessed philistine but I was not alone in rejecting at once the myth of Mme. Bernhardt's histrionic ability. As everybody knows, almost no one who is not a Master of Arts and does not live in the cheese and wine belt of Hampstead can gain admission to the National Film Theatre. Yet when this piece of film was shown the entire house choked with mirth. A man sitting near me became so convulsed with merriment that he fell out of his seat into the aisle and lay there with his feet waving in the air like a supine wood louse.

From all this we may conclude that it was not by means of what she did on the stage that Mme. Bernhardt ruled France. She brought the nation to its knees by what she did with the rest of her time—by spending her nights in her coffin and her days surrounded by a cheetah, a monkey, a parrot, four dogs, and M. Gustave Doré.

The same writer who accused Mme. Bernhardt of lacking humor praised her autobiography for having that very quality. Perhaps in this apparent contradiction lies one of the secrets of life-style. An element of self-mockery should always be woven into our attitude toward and our description of our past behavior but should never prevent us from flinging ourselves with total commitment into the present and making it a backcloth for our most extravagant gesture.

When we turn our eyes toward the movies, our pulses quicken; the shadows on the landscape grow sharper; the background music swells. By these and other traditional signs we are made aware that we are approaching the spot where the treasure lies buried. We have entered the region where once the sacred mushroom of personality grew as tall as the sequoias.

The movies have been through many phases. In the beginning quite naturally audiences felt only amazement that such an invention existed. It was not until Mr. Cecil B. de Mille arrived in California that people stopped merely staring at the pictures and started looking through the screen aperture into a world of dreams. Even then it was the story that mattered most. However basic it was, because it was being told in an unfamiliar medium it was difficult to follow and demanded total attention. I myself did not begin going to the pictures for about ten years after this phase began. By then, in the early twenties, stars such as Miss Pauline Frederick were already claiming more public notice than the narrative. Mr. D. W. Griffith wished to prevent this shift of emphasis. The way Miss Lillian Gish, with her unconquerable tact, tells it, this seems like a miscalculated public relations job.

As I see it, like a true stylist the creator was doing everything in his power to avoid being eclipsed by his cast. He failed. His public was simply not sophisticated enough to take in the idea of a director's film. This concept was not popularized for another twenty years. Then the fair name of Mr. Alfred Hitchcock began to be printed in the credits of his films as large as those of his leading players. In Mr. Griffith's time people were absolutely unaware of the technical devices that he was inventing at the rate of almost one a day. (Even the word "film," which is basically a more technical term than the word "picture," was hardly ever used.) The camera was the audience's eye, and with it all they wished to do was to gorge themselves upon every aspect of the divine being.

I came from a middle-class family living in Sutton, which is a suburb twelve miles outside London, in Surrey, and because I went to my first movies with my mother it was to some extent with her limited vision that I saw them. (I don't remember my father ever going inside a cinema.) I was in a secret trance while each episode of *The Phantom Rider* lasted but afterward I pretended to think it ludicrous. One of the great jokes at home was to clasp one's hands and roll one's eyes as actors did on the screen. In my mother's view the pictures were all right for children and the lower classes but they were trivial and, worse, they were American. For people of refinement there was only the theater.

I was at least seventeen before I was often with people who talked without condescension about the movies. My professor of literature said, "If you want to understand Shakespeare, go to the cinema." Even so, I and most of the rest of his students interpreted him to mean that we were to go to culture films (German) rather than entertainment ones (American).

Thus it was that more than a decade of moviemania passed me by. I am amazed at this, for it is impossible to exaggerate the influence that the cinema had on urban life during the last phase of the silent movies and the beginning of the era of talkies.

Apparently, from the industry's own point of view, the peak years were in 1946 and 1947, but even if more seats were sold during those years, people occupied them less wholly. By then Londoners had all played at least bit parts in their very own Technicolor wide-screen production—the war (though many complained that it was less moving than the silent version released twenty years earlier). They now went to the pictures merely to relax. During the late twenties and early thirties they had gone to worship. That was when so many cathedral-sized cinemas loomed up in Edgware Road that they seemed to stand shoulder to neo-Egyptian shoulder.

In those darkened temples people lived the vivid hours of their lives; the dreary residue endured outside passed like a trance. It is useless to ask which level of experience was the more real. Reality is only the dream in which our enemies believe.

In modern times our imaginations are enthralled by less poetic myths. We believe in statistics. Among other things these tell us that movies are made for a girl between the ages of fifteen and twenty—preferably with an unhappy love life. This piece of information would not formerly have been much help. The movie industry operates in the realm of the affections and there, until a few years ago, all women were fifteen.

The term "a woman's film" is fairly new not because an opportunity to weep is a recently discovered pleasure but because at one time there was no other kind of movie. Although stars were advertised as having S.A. or IT or

OOMPH, it was never with the hope of luring men into
the cinema. The portion of the male population that
went to the pictures did so in order to sit with their girls in
the dark. The women went, not for a reciprocal fumble
but to gather material for the dream of which these
smash-and-grab raids on the part of their menfolk
deprived them.

Sex may have already come to real life but comfortable
sex was still a long way off. Proof of this lies in the look of
the poorer districts of London at night. They are occupied, if at all, only by people on their way somewhere
else. During the dreaming years, there was not a dark
doorway between King's Cross and Paddington that did
not partly conceal a couple that for politeness' sake we
will describe as courting.

Faintly bawdy songs about film stars were well known,
but I never heard a man express any real desire for any of
the divine women. The great female stars were wor-
shiped by girls because in their looks, in the tone of their
voices must lie the key to the Pentagon of sex—that
stronghold from which all men could be ruled and life
(meaning sex) could be enjoyed on a woman's terms.

The films of that era were truly immoral. They were
governed by strict—almost ludicrously strict—rules
about what areas of human skin could be exposed to the
cameras, about how long a kiss on the lips should last, and
about how much of a man's body must remain off the bed
while he was attempting rape. The stories, where after all
immorality must always reside, were subject to far less
careful supervision. Almost every vehicle for a female

star carried the same message: You can have your sexual cake and eat it. Since these films were shown to total believers at a time when, in fact, virginity before marriage was still all the rage, they were, to say the least, misleading.

Thickly unmade-up, the leading lady would say farewell to her good, kind parents and set off for the wicked city. She had hardly taken her seat in the train before Mr. Lionel Atwill was asking, "What is a first-class girl like you doing in a third-class compartment?" From that moment onward it was down, down, down she went while her dressmaker's bills rose higher and higher. After hours of degradation, her soul sick of sin and her skeleton bowed by the sheer weight of fur and precious metal, she took refuge in the arms of any one of the industry's stuffies—Mr. Clive Brook or Mr. Herbert Marshall.

As each new star went through these rituals of fall and redemption, she was packaged by Mr. Louis B. Mayer and his merry men as holding the new, improved stratagem for vanquishing men. Since the movies were not run by scientists but by abominable showmen, every discovery was put forward as the final answer.

The posters were hardly dry on the billboards before a thunderous din like the sound track of *Red River* was heard throughout the island. It was caused by the hooves of thousands of women as they rushed upstairs to their bedrooms to throw out the aids to the now obsolete image and get to work with the new.

In the late twenties the streets of London were entirely overrun with near-Garbos; in the early thirties by failed

Dietrichs. Before Miss Veronica Lake was finally drained, thousands of girls had been dragged by their hair into looms, through mangles, and under presses as though they were living in a mechanized Aztec kingdom, merely because this particular actress wore her hair hanging over one eye.

Wherever two or three women were gathered together—in factories, in kitchens, in laundries—from dawn till day's end the conversation consisted of each of them acting out the entire story of the current release to all her workmates—with actions if there was room.

Of course the girls in the balcony seats never did learn the secret of style because, in their haste, they neglected to do their homework. Between visits to movie palaces, they indulged in day-long fantasies instead of submitting themselves to limited periods of gruelling self-examination. It is a pity, for they stood at the entrance to the richest mine of style ever opened.

What went wrong? Once again life and art began that fatal movement toward each other. Whenever this occurs, it must not be blamed on the philistines alone. It is often they who would like art to remain pure and free from the compromises of the world, and when they do wish for a convergence of art and life it is usually because they want life to be more stylish. In the case of the movies, in their faltering way the devotees tried to become as beautiful as their idols and life has abetted them. Wealth, leisure, sex, and false eyelashes have gradually been brought within reach of all. This seems to me a good thing. What is bad is the movement from the other side.

Artists never accept for long the limitations of their medium. They never seem aware that these are what gives their art its style.

The decline of art is always from realism to naturalism. Realism is usually a technique of subtraction by which an artist lays bare the essence of his subject; naturalism is always a trick of addition with which a charlatan seeks to give conviction to a lie. In the art of moviemaking this disintegration was a particularly sad story because, in the beginning, so much ground lay between it and real life—literally. Now pictures have even lost their national style and are made wherever money is available, but once they were almost all made in a desert on the other side of the world. In that paradise there were no buildings—only sets; no people—only actors. From the dawn of Miss Mary Pickford to the end of a perfect Miss Doris Day, not one sincere word was spoken. Even when sound came it brought with it an unexpected fringe benefit. At first its use presented such technical difficulties that the whole industry moved indoors. Then not only the cast but every stick and stone surrounding it was artificial.

This happy state of affairs could not last. Talkies moved out into the streets and even began to boast that they were doing so. At the same time, as with stage stars, movie actresses allowed it to be known that they were human. During the reign of Mr. Mayer, truth about players under contract to him came only from his lips. Not only their faces but even their lives were made up. This kept their life-style all of a piece. Now, even if the human race still had a capacity for wonder, it could not lavish it upon film actresses. We know their real names; we know their real measurements; and when magazines like *Confidential* still flourished, we knew or thought we knew about their bizarre bedroom antics. It is true that even after it became known that Mr. Zeus was a sex maniac he remained a god, but few people have a public relations department like his. The stars lost their glitter and decadence set in. Audiences then began to take an interest in the movies themselves—not in the stories they told but in the manner of telling them. Instead of rejoicing in the triumph of virtue, they got enthusiastic about accurate post-synchronization and skillful back-projection.

When the movies abandoned the star system they did not lose everything but gave up what was uniquely theirs. No other medium can so fully deploy physical style. No other art form is at once so intimate and so distant. The cinema allowed an audience to spy on its heroine as though from a nearby window with powerful binoculars. The subject appeared not to know she was being watched. She did not pause for laughs nor raise her voice above the sound of coughing in the auditorium as mere stage actresses do. She continued her private life undisturbed, thus giving us the exciting feeling that there was nothing, however personal, that she might not do. Yet we could scrutinize her so closely that we saw her skin texture and watched the beating of her heart. This combination of opposing qualities made every film into an overheard scandal, compared with which a play, performed with semaphore gestures and spoken with tub thumper's diction, is as intimate as a parade ground command.

We knew the stars as we know God. They were all-pervading but remote; we created them but their fate was not our fate and they remained intractable, unlikely to respond to our prayers.

Movies were not made in sequence. The whole purpose of this stratagem was to save money—to make full use of one set before striking it—but unintentionally it conferred on the final product an extra benefit. Actresses could survive this patchwork method of manufacture, could give their performances any degree of consistency, only by perfecting a set of idiosyncrasies. The more predictable these gestures came to be, the more inevitable rightness the character acquired. What audiences went to see was not an act of impersonation but precisely this triumph of cast-iron personality over the hazards of life, desertion, betrayal, misunderstanding, and death.

When the tide of popular taste turned, it was the indestructability of the stars of the twenties and thirties, far more than their exaggerated make-up or their unfashionable clothes, that audiences laughed to scorn. Their mockery was misplaced. These films embodied an eternal truth. Not merely on the screen but in life itself, the only way in which we can survive and be seen to survive is by the dexterous use of a carefully selected and patiently rehearsed set of mannerisms.

As long as the star system sold pictures, style was worth its glitter in gold. No one is born stylish but some have style thrust upon them. Those actresses who began their careers with looks but no idiom were given one by their agents, their sponsors, or their directors. Sometimes the synthesized public image did not fit the woman on whom it had been imposed. In that case she either rebelled like Miss Davis or acquiesced uneasily like Miss Jean Harlow. The girls who accepted defeat on this issue suffered all the irritation that besets people with broken limbs in plaster casts. Stars compelled to wear ill-tailored life-styles were not always able to shrug them off. They came to have the unpleasant feeling that they were receiving fan mail addressed to someone else of the same name. They seemed to have the world at their feet but not in their hands. For this reason, in every envelope containing a long-term contract issued in Hollywood there was also a small machine for cutting wrists.

Suicide can either be a gesture of frustration made by the victim of an unsuitable life-style or it may be the last graceful gesture of someone whose style has been completely mastered. I am sorry to say that Miss Marilyn Monroe's death seems to me to fall into the first category. She was not able to make full use of her fame and thus became its victim. She did not become a stylist.

Sometimes, if everything was just right, it was an advantage than an actress could be seen to be operating in a style that had been dictated to her. We are speaking of an age when, though people of the movie-going level of society had far less spare time, they were willing to take more trouble with their happiness. As long as a star's image was steady, it did not matter that it was cloudy.

Indeed, this lured them forward to peer and to specu-
late. The object of their interest became someone whose
behavior they could calculate but whose motives could
not be wholly understood. This imparted glamour,
which is not mere beauty, exists wherever all is promised
but not all is given.

The most perfect example of a star with this hypno-
tized and hypnotic quality was Miss Brigitte Helm. Perhaps inevitably, soon after the advent of the talkies she
became extinct. Though she would now be no more than
sixty-seven or eight and was certainly the most beautiful
woman that human eyes have seen, she is never photographed on the yachts of millionaires; she never appears
in cabaret in Las Vegas; she does not write her memoirs
nor does she cause them to be written by Mr. John
Bainbridge. At the National Film Theatre, where nothing is too archaic to be screened, the only film in which
she starred that is ever shown is *Metropolis*. Even the
critic, Mr. Kevin Brownlow, who is besotted with the
silent movies, hardly mentions her name, whereas one
would have expected to find him rolling about on the
carpet in front of one of her stills. It is possible, even
likely, that before he left for the Islands of the Blessed
her director, Mr. Fritz Lang, dismantled her. He certainly assembled her in the first place.

The program notes for *Metropolis* told the reader that
Miss Helm was eighteen when Mr. Lang chose her for the
leading part and that, during the entire time it took to
make the film, she lived in his house with him and his
wife so that the world might not touch her and so that she

could come under no other influence than his. It hardly
matters whether this piece of information is true or
merely a nice chunk of public relations hokum. To see
the film is to feel that it is true. Its star has the quality of a
walking, talking, living doll. This impression is rein-
forced by the fact that the story shows her being manufactured out of scrap aluminium by a wicked scientist. To
this cold, mechanical style she forever after clung
through thick make-up and thin plots. Though subsequently directed by other moviemakers with slightly
different styles, she continued to make very few conces-
sions to real life. However mundane the events described
in the scenarios, she lifted every film onto the love-and-
death level, went shopping as though to the scaffold and
drank tea as though it were a love potion.

Sad to say, even an audience riddled with culture
would now have to watch one of her films with two pairs
of eyes. It could not fail to admire her monumental
beauty and the unequaled economy of gesture with
which she performed every action. At the same time it is
impossible to deny that the whole thing is a ludicrously
trumped-up charade made worse by beehive hats and
six-yard trains hanging behind skirts that hardly cover
the divine knees.

Miss Helm's personal style was a peak rising above the
great range of her national style. Germans evidently lean
toward the cold, the cruel, the doomed. If they do not
and it was only the head of UFA films who admired these
qualities, then he must have held an entire nation in his
iron grip like a celluloid Hitler, for every picture that

came out of Germany during the twenties was of this kind—*The Cabinet of Dr. Caligari, Warning Shadows, The Student of Prague.*

When, because the talkies raised a barrier of language between us and culture, we were compelled to turn our attention to entertainment, the change of atmosphere was like that which we feel when we come out of a cave on to a sunny beach. Compared with the icy somnambulists of German films, Miss Garbo seems wide awake—almost human. She too had her private hypnotist in Mr. Mauritz Stiller, who took her to America, but the remoteness with which he invested her seemed to come not from having no heart but, on the contrary, from being too sensitive to the tragic aspect of love.

A critic has said of Miss Garbo that in her pictures she always seemed to be the only member of the cast who had read the script to the bitter end. She was the only one who knew that it would all turn out badly. One might go further and say that tragedy was her element. The trashy material of most of her films was brought to life only when the moment arrived for her to deliver one of her lectures on hopeless love. This appeared to be the one subject which, though it is unlikely that she ever experienced it, appealed strongly to this actress's imagination. The voice for which everyone in the world—except one or two mean-spirited movie reviewers—had been waiting turned out, in Garbo's talkies, to be permanently charged with emotion. It was made for the throaty crooning of captionese. "I am so sorry for all the women that are not in love"(Inspiration). "To me love is only a little warmth in all this cold" (Romance). And, of course, the greatest caption of all time: "Perhaps it is better that I live only in your heart, where nothing can stain our love." How heart-rending that would have looked in Caslon Old Face type, white on black, with a drawing of a camellia in the bottom right-hand corner!

To realize fully the magnitude of this woman's life-style we need only imagine how empty—indeed, how embarrassing—these words would sound issuing from the lips of any other movie actress. Audiences left her films in a trance of numbed acceptance, wondering vaguely why the world, which to the rest of humanity seemed such a jolly place, was not good enough for her.

I am warned that Sweden resents the opinion held by the rest of the world that suicide is its national style. It is a pity, for without it the country has no style at all. Nonetheless, to avoid causing the withdrawal of ambassadors, I will not put forward the suggestion that Miss Garbo's gloom was typically Swedish. We will consider her not geographically but in relation to history.

She came to California when the reign of Miss Gish appeared to be ended. Recent events have shown us that in fact this lady's reign never ends. She is a true exponent of life-style and has converted the career in which she set out into the profession of being.

Judged by the limited standard of Hollywood, however, her kingdom seemed to diminish. She represented sweetness and light, and from these two commodities the consumer appeal was passing rapidly. Someone had to be discovered who could win the love of several men in one film without in real life losing that of Mr. Mayer. Miss Garbo did this by fornicating liberally but with massive distaste. The love behind her was always carnal and revolting, the love ahead divine.

In *Anna Christie* she played the part of a prostitute and yet looked forward, if not with relish at least with shakily founded hope to being married to Mr. Charles Bickford. To modern moralists the argument of this film would seem perverse. Mr. Bickford agreed to accept Miss Christie only after she swore that she had never loved any of her customers in St. Paul. Nowadays this assertion would make disgusting a way of life that might otherwise be thought of as an interesting experience.

Even so Miss Garbo was the harbinger of the permissive society. Whatever happens to sex from now on, there will be no more divine women. Some clot left the greenhouse door open.

But before this happened, there emerged from the middle of Berlin someone who was prepared to carry the flickering torch once more around the arena. Her name was Miss Marlene Dietrich.

Unlike the other stars that we have mentioned, this lady admits—nay, boasts of—the influence upon her career of a man. His name was Mr. Josef von Sternberg, and so pronounced was his Pygmalion-Galatea relationship with Miss Dietrich that he was called Svengali Joe. In his book *Fun in a Chinese Laundry*, though every other page is wet with the author's tears at people's ingratitude, there is one that is scorched with rage at his Trilby's openness on this subject. Nevertheless he does not deny that she was his invention. From this we learn, though the knowledge is of no earthly use, that while beauty in art is simply the pattern made by means that are adapted to an end, in life it is not a woman but a man's idea of a woman.

In the case of Miss Dietrich, the permissiveness that has now engulfed civilization went a step nearer the precipice. It was only natural that it should. She arrived later than Miss Garbo, stayed longer, and came from a country where the existence of sin is at least acknowledged. And the morality of her films would not have been subjected to quite so much scrutiny.

She was not a Goldwyn property. When she appeared in *Shanghai Express* she played the part of a white (or scarlet) woman drumming up yellow trade. Critics, claiming to be "reasonably broad-minded," were outraged. This was partly because in the eyes of the elders she had sunk lower, but also because she patently did not share Miss Christie's remorse for her past. She did her hair five different ways in one train journey and gave every impression of doing with great aplomb what she knew very well how to do. The sphinxlike quality that Mr. von Sternberg imparted to her differed from that of other female stars. It was neither unearthly nor tragic. It was mocking. It held her within the great tradition, allowing audiences to read into her mask whatever secret yearnings they wished and to garland her neck with their dream as though she were a visitor to Polynesia, but this particular enigma variation was also an expedient.

When Miss Dietrich first arrived in California she was less sure of the American language than she later became. It was therefore best for her to murmur her veiled way through various captions with as little passion as possible. Most of her strength seemed to go into the raising and lowering of her eyelashes. No wonder. They were at least an inch long.

By the time she finally ended her association with her mentor, her image was faultless. Quite rightly she never made any drastic alterations even to her lighting. She feared to lose the love of Paramount—even of the whole world. To save her fabulous face and nod distantly to changing fashions, she extended the immaculate mockery that she habitually lavished on men to include the movie industry itself.

In a film called *The Café of Seven Sinners* (Miss Helm had once appeared in a piece called *The Yacht of the Seven Sins*) Miss Dietrich played a woman so wicked that even her street clothes were backless. Mr. Broderick Crawford had to pummel her into agreeing not to drag the fair name of Mr. John Wayne in the dust by marrying him. Convinced, she sailed away from Singapore to higher heels and thicker maribou. In the last moments of this picture we see her asking the ship's doctor for a cigarette. "You're a real pal, Doc," she says. This line is clearly not a caption and another actress might have accompanied the words with a playful punch on the biceps, but not the imperturbable Miss D. She murmurs the words languorously, as though she were in bed with the doctor, and turns toward him a face half-shrouded by a diagonal eye veil. The lashes of one eye poke through the black net like arms of mute appeal; in the other is the famous equivocal expression. In spite of the jarring American slang, no breech of style has been permitted. Glamour *omnia vincit.*

As M. Cocteau warned us, behind the gods there are other gods.

The lure of these three actresses and of the whole aviary in which they were the most exotic specimens was that they could be seen to establish, apparently with little or no effort, a mysterious ascendancy over men. Since there is now hardly any difference between the sexes, the idea of male domination is rapidly passing. Women have decided to conquer by protest a world they failed to win by seduction. Consequently the sight of any kind of exploitation of femininity brings back nothing but the humiliating memory of previous defeat. Not only long cigarette holders and slave bangles but the very idea of a vamp has turned out to be a passing fashion.

With whatever affection we may look back on the movies of Miss Dietrich, we must never forget that the film industry has produced some personalities who were beyond time. These fixed stars had few or no imitators other than stage mimics and yet (perhaps I should write "and therefore") they were greater in terms of style than any of the players we have mentioned so far. Mr. Harpo Marx and Signorita Carmen Miranda, for instance, were people we could not have invented if we had sat up all night. They embodied nobody's dream but their own; they were what the writer Mr. Philip O'Connor would call autofacts.

The greatest autofact of all was Miss Mae West. She claims that through the long, dark years of the depres-

sion she upheld the movie industry single-handed—
though, as she would undoubtedly say, hands had nothing to do with it.

When her films were first shown in England, they were
publicized as bringing new hope to women who might
otherwise have given up. This doesn't seem very likely.
She could be imitated (and frequently was) by men in
drag, but no woman would have had the nerve to
seriously attempt to convert to her own private use such
luxurious self-glorification, nor would any man I can
think of have felt able to cope with her demands. There
was in all her work an element of self-mockery.

Self-mockery is a technique that all of us may as well
learn. Individuality is feared by those who have none.
Therefore a stylist must expect to be viewed if not with
terror then with contempt, which is to fear what glucose
is to sugar. This does not mean that anyone should ever
sink so low as to disparage himself in the presence of his
enemies, but he may indulge his style with such obvious
relish that a part of his personality detaches itself from
the rest and sits with the audience sharing at least the
general incredulity at the behavior of the actor. This,
according to Sir Max Beerbohm, is what Miss Bernhardt
always did. Even in her most terrific moments, he tells us,
one half of her soul was in the position of a spectator. No
one is obliged to ingratiate himself with his audience, but,
should he wish to do so, this is one way in which it can be
done without loss of dignity.

A lot of actors and actresses are said by critics to display
a gift for self-mockery when really they are doing no such

thing. Miss Judy Garland, for instance, was ready at any moment to black out a tooth or put on fifteen-inch shoes. This kind of thing was not true self-mockery. It never made her audiences think of her as a freak. They knew that she and the character she was playing remained their sweet composite self beneath the grotesque disguise. Indeed, for the moment she became all the sweeter because her antics appeared to be desperate measures to which her need to hear our laughter had driven her. She never for a second made fun of her pathological greed for fame. Really what she was doing was clowning. Because pathos was her life-style the world forgave her, but in general clowns are a detestable subhuman race forever raucously inviting us to laugh at accidents befalling ill-favored victims.

Miss West was never guilty of anything so styleless as slapstick. Self-defilement was not for her. On the contrary she exuded self-gratification as hemp flowers secrete resin. The most we can say is that this essence was occasionally spiced with a pinch of wit, as when in *Go West, Young Man,* wearing a skin-tight black dress, she walked down a garden path behind a huge black sow. The beery breath of clowning would have tarnished her image.

This was seen to be so when finally she was cast opposite Mr. W. C. Fields. Though he was not truly a clown, he had such a harsh personality that, in its chilling air, the orchid of Miss West's style could not blossom. The film was a failure. This disaster may have been an instance of what film historians will one day come to call

Brooks's first law. When interviewed in New York by a member of the staff of *Sight and Sound,* Miss Louise Brooks explained that, if the front office began, like Dr. Frankenstein, to fear the power of one of its creations, someone was sent to persuade the star in question to make "a real stinker," so that she could then be killed off without too loud a cry of anguish from the fans. Even if, in Hollywood's estimation, Miss Brooks is no longer alive, no one can say she is not still kicking.

Until the making of *My Little Chickadee*, Miss West had it all her own way. She played opposite men as handsome as Mr. Cary Grant and as vast as Mr. Victor McLaglen. She was given stories in which, after a prolonged build-up, she undulated through saloon brawls, police gun-fights and even Salvation Army jam sessions without bursting a single seam of her female impersonation clothes. This was surprising, for she looked as though she had grown up in them—and she had grown up. Also, whatever disagreements she had with her rivals, they were never allowed to derange her coiffure, which looked as though it was made out of boiled snow. She spoke her own lines and even the way they were present-ed was her own invention. At a period in movie history when the director's word was law (to everyone but Miss Davis), Miss West insisted that her jokes be followed by a slow fade-out. The traditional quick cut would have deprived her of a slow, triumphal departure from the battlefield and the audience of the sight of her bustle preening itself with pride.

Of all the stars that ever shone in the long California

night, Miss West came nearest to breaking the celluloid barrier. Although in each of her pictures she played a part—that is to say, she was never actually addressed by the other players as Miss West—she came nearer than anyone before her to looking into the camera aperture to ask her audience the direct (but rhetorical) question, "How'm I doin'?"

Almost all star roles during the dreaming years were self-glorifying, but usually the actors or actresses who played them believed utterly in the bravery or the virtue by which in the story they conquered—and they sought to convince their devotees. When Mr. Alan Ladd used regularly to shoot down whole bowling alleys of baddies without even turning down the collar of his raincoat, he was sincerely trying to bring a message of hope to tiny men. Watching Miss West was different. What audiences enjoyed was not so much the triumph of a fictitous character over crooked cops and two-timing Romeos as the sight of an actress hilariously trouncing the entire movie industry. At long last we saw the female dragon rise to gobble up the ridiculously armored mythmaker and then casually brush the crumbs from her ample bosom.

Although in their heyday the movies were a machine uniquely equipped for packaging and marketing style, an aspiring stylist would have been well advised not to go to Hollywood without a return ticket. The price of a cement footprint was—and still is—considerable.

At six in the morning a movie actress is roused from sedated sleep and rushed to the factory where she is temporarily employed. There a swarm of workers buzzing with envy prod her into a dress with an eighteen-inch waist and smear her face with the expression it used to have in the days when Mr. Darryl Zanuck liked her.

Then they lead her into the light blinking like a suspected person brought in for questioning. Hour after hour, obeying megaphoned instructions, she turns, smiles, says "I love you," turns, smiles, says, "I love you," turns, smiles . . . until at dusk she is dismantled and taken home.

This doesn't sound much like a day in the life of somebody loved by half the world, but once it well might have been.

Furthermore, there are other drawbacks to a film career than mere physical discomfort. The movies are part—by far the greatest part—of the entertainment business but, in a sense, they are not a doing but a making profession. A stage actress says, "I am doing a new play," but a movie star admits she is only making another picture. As with the work of a painter, so it is with a movie. It is a thing and it is taken away from the people who made it and shown in another place to strangers. Very few stars have any say in the way their work is distributed and not many have a direct financial interest in its success.

The very geographical isolation that gave the movies their former perfection exacerbated to a slight extent the malaise from which film folk often suffered. It gave them the sensation of being exiles. The same shudder that ran down the spines of moviegoers when they emerged from the cinema into the cruel light of the street also shook the stars as they left the studios on the last day of shooting. The crew turned its attention to another actress and the fans cheered their favorites on another side of the planet. The object of all this emotional reaction sat in a bizarre house in Beverly Hills anxiously reading scripts as similar as possible to the one she had lately illumined with her luster. A sense of anticlimax went with the territory. A feeling of missed reward grew without sustenance, like the cactus that is indigenous to the state, until it flowered into a conviction of persecution. The vast salaries were forgotten. Sometimes, because they were not the whole answer, they came to be regarded as an insult. The girl who had all sorts of plans for doing something sensible with her wages turned into the star who threw her money about contemptuously. Many women appear avaricious but few of them love money as men can with a pure flame. For a lot of men banks are holy places and the accouterments of finance—checkbooks, cash boxes— acquire the magic of fetishes. A stockbroker's sleep is not made lurid by visions of overfilled bras; he dreams of bulging yet uncrackable safes. His prayers are all in numerals.

Women are a baser breed. Before they can enjoy riches they must convert them into things: houses, furs or—as in the case of Miss Paulette Goddard—diamonds. Sometimes, having in youth preserved their vital statistics from damage by parturition, they later on send out for store-bought children only to find that these are not like the dolls dressed in national costume that used to adorn the couches of prostitutes but are living organisms constitutionally free to become horrible presidents. From these and other similar causes is born the Judy Garland syndrome.

Of those who survive both stardom and the loss of it, the majority depart from dreamville leaving not a trace behind (whatever happened to Miss Corinne Griffith?). These are the stars whose style, however luxuriant, was not hardy enough to survive the winter of Mr. Mayer's discontent. They never managed to convert a career in the movies into the full-time profession of being.

This hardly applies to Miss Davis. When asked by a reporter if she would make any more films, she replied, "I may have to." If this is true and only material need has driven her to make her recent movies it hardly shows. Even in earlier pictures, where she played sympathetic parts, her contempt for the film industry as a whole and possibly for life itself seemed to beacon the world's night. Her ferocious acting in pseudo-horror films looks less like a loss of prestige than a vindication of her original judgment.

In spite of all this, the kiss of Hollywood must not be diagnosed as fatal. There are actors and actresses, not necessarily as great as some of the names we have mentioned, who are not merely alive today but who, when they left California, brought their style with them absolutely intact. When these people tunneled their way out, they came up in very strange places. Mr. Ronald Reagan emerged into domestic politics and Miss Myrna Loy into world diplomacy. Miss Gish has operated in the

lecture hall while Miss Garbo, wearing the most conspicuous disguise the world has even seen, inhabits a sort
of free-range nunnery. We pause before her shrine, a
hush falls upon our trivial conversation at the mention of
her name; but it is impossible to shake off the fear that
she is nothing but her style. To complain of this sounds
like blasphemy, but if it is true, then something is wrong.
Whatever our thesis appears to be, ultimately we are
always speaking of happiness. We must not only invent,
polish, and project a life-style; we must also enjoy it.

Coincidence is the triumph of style over chance. Over
and over again the stories of Miss Crawford's pictures
brought their heroine from straitened—sometimes even
sordid—beginnings, not so much to happiness as to social
vindication. Enjoyment hardly seemed to enter into
consideration. Whatever reward Miss Crawford's screen
self finally won, the cost always seemed to have been too
great. This impression was given less by the script than by
the star's way of acting. Her performance seldom contained any of those happy incidents that decorate the
work of people like Miss Dorothy Maguire. Audiences
felt that what Miss Hedda Hopper said of Miss Crawford
was true. She had nagged herself into becoming a
competent actress. We have been told that before her
triumph in *Mildred Pierce* Miss Crawford was having
trouble with her career, but even if this information had
never reached our ears we should still have thought of
her as hacking her way through script after stubborn
script. The difficulties the heroines that she played had
in subduing the women's watch committee of some

American provincial town seemed to be one with the rigors of staying on top in the movie business. To me this makes Miss Crawford a supreme stylist. What in a lesser person might have been seen as a defect she turned into a kind of glory. She converted what may have been a real resentment of people who wanted less or didn't have to fight so hard into a unique personal idiom. She was the living example of one of the basic rules of style: That which cannot be wholly concealed should be deliberately displayed.

Miss Crawford's escape route under the barbed wire of Hollywood surfaced in the boardroom of Pepsi-Cola. This was an outcome that no one would have dared to predict but, after it occurred, it seemed inevitable. It was an actual event exactly like the final reel of one of her pictures. By means of it Miss Crawford moved from a doing profession into the world of being. To the lovers it may have seemed an ecstatic accident that Mr. Alfred Steele met and fell in love with the movie star. Really it was fate. There had to be a person—there had to be an event however tragic that would open a door into the corridor of industrial power.

Whatever wealth this marriage brought to her is absolutely beside the point. She had been earning what most of us would call big money since *Our Dancing Daughters* which was made when even I was young. What matters is that she had been placed in a position of power where she could never again be discomfited by society or by Mr. Sidney Greenstreet.

This perfect alignment of art and life ought to have

been of benefit to both. However dull they may have
been before Mrs. Steele enlivened them with her presence, Pepsi board meetings must surely have acquired all
the glamour of a Technicolor, wide-screen first feature
while, doubtless, the real situation gave to the star's
performance the kiss—or rather the bite—of life.

While being fêted at the National Film Theatre in
London, Miss Crawford was asked why she had never
gone on the stage. "Because then," she replied, "I would
be as nervous all the time as I am now." If she was scared,
it certainly didn't show. She seemed incandescent with
self-assurance. When, for this public appearance, she
arrived at the theater, I saw Miss Crawford in real life. A
car as long as the Thames drew up outside the cinema
and from it stepped those two now famous children.
They stood in the lobby looking bewildered—and no
wonder, for after two or three minutes Miss Crawford
alighted from the same car and kissed the two of them as
though she hadn't seen them for months. I was standing
half-way up a short flight of stairs to get a good view. She
looked at me for only a moment before her hosts took her
attention, but when she had gone through the stage door
I felt impelled to turn around and see if my silhouette
had been burned on the wall behind me as were the
images of certain victims of Hiroshima.

Another star who has made a similarly seamless transition from the movies to real life is Miss Dietrich. She
differs from Miss Crawford in that she was born not to
rule but to love. When asked at a press interview why the
tender passion took up so much of her time, she replied

with another question, "What else is there?"

Naturally, holding these views, she did not leave Hollywood for big business but for the halls. In her films she constantly played the part of a woman who became a cabaret singer (*Blonde Venus, Song of Songs, Seven Sinners*). Nothing has changed except that we now sit where once the cameras stood. Strictly speaking the form of entertainment that she has chosen to give us went out of fashion when her movie career was at its height. It is a tribute to the force of Miss Dietrich's personality that she has been able to reverse for a while the laws of time and bring vaudeville back to life. In her act she wears pretty clothes, smiles, and sings, just as Miss Ida Barr and Miss Nora Bayes did long ago.

Miss Dietrich has made so bold as to state publicly that she *never* liked acting for the screen. Among the objections she lists is having to sit still every morning while, under the instructions of a continuity girl, a dresser placed each separate hair of the star's head back into the position it had occupied yesterday evening when the cameras stopped. While understanding this grievance, we can't help feeling that a deeper cause for grievance is the fact that movies put a film (literally) between the lady and her admirers.

Proof of this can easily be found by observing the stage performance. Miss Dietrich hardly sings at all. From buttered lips she breathes a few idiotic moans which pencil in the tune as written. What brings the audience and more particularly the performer to life is the applause at the evening's end. To public acclamation she responds with every feminine reaction possible outside the bedroom. Indeed, the unique quality of her show is that it is almost totally sexless but wholly loving. As long as the curtain goes up and down for her curtain calls, she is by turns humble, teasing, happy, moved. At one moment she marches triumphantly through the center opening of the curtain; at another she seeps around the side of the proscenium; one bouquet she will receive diffidently, another eagerly. After a while it is Miss Dietrich's reception of our applause that we cheer, and in return we ourselves are cheered at seeing time defeated by what one might call a probity of style.

All these movie people who now rule the stage or big business or politics do so not merely because their fame provides a sort of tourist attraction but because the style they learned through the disciplines of acting for the movies has made them, in any sphere of activity, invincible.

BEFORE THE END

Once we have mastered the gestures, the voice, and the words that will express our chosen image of ourselves, we are prepared to leave that place that has so far been our hermit's cave, our athlete's gymnasium, our actor's sound booth. Having progressed from making to doing to being, we are ready to move out into the world.

If sophistication is a matter of being in control of our primary reactions, we may now be sophisticated. At least we shall be fairly confident of ourselves and may, with any luck, be confident of others. Our object will be to enjoy ourselves, but to make sure that our names are permanently on the cast list, it will be advisable to be of interest to others. This aim must never be confused with the desire to be popular.

We must never be rude unintentionally and very seldom by design, but we must accept right from the beginning of our career as a stylist that a clearly marked personality cannot be universally liked. If offense is given it may lead to our being struck off one person's visiting list, but this will almost certainly lead to our being welcomed elsewhere.

The whole question of a stylist's relations with society is very subtle. Style is its own reward but the fringe benefits of fame, money, and power, which spring from the public projection of style, are so great that at times it is difficult not to write as though the winning of them were the real object of style. This is not so. Other people are a mistake, or if that seems too sweeping a statement, concern for others is a mistake. This relationship between a stylist and his public is usually mutual. Very few people bother about the welfare of anyone with a highly developed style. This is an aspect of Mr. E.M. Forster's civilized apathy.

Style is a shield; style is a sword; style is a crown; and style is also an automatic invitation card to the party at the end of the world.

When Miss Elizabeth Taylor was Mrs. Mike Todd, her husband gave a party in her honor at Madison Square Garden, whose marquee that evening bore the unusual legend, "A Private Party Tonight." Of this occasion Miss Taylor later said, "It was a shambles, when all we'd wanted was a cosy get-together with eighteen thousand of our closest friends."

Perhaps on that one evening this matchless stylist forgot one of the rules of style: Never attend a gathering so large that everyone in it cannot be made fully aware of your presence, and do not mix too often with those who cannot appreciate its quality.

It is not true that stylists surround themselves with dullards. Conversation stimulates conversation; wit provokes wit. Every hostess knows no party can be called a success until everyone is talking and no one is listening.

In the midst of this pageant of personality, this hail of epigrams, who will hear the bang or the whimper when it comes?

Who will care?